About this book

This workbook contains practice to support your learning in P1 & P2 maths.

Questions split into three levels of increasing difficulty – Challenge 1, Challenge 2 and Challenge 3 – to aid progress.

Symbol to highlight questions that test problem-solving skills.

Total marks boxes for each challenge and topic.

'How am I doing?' checks for self-evaluation.

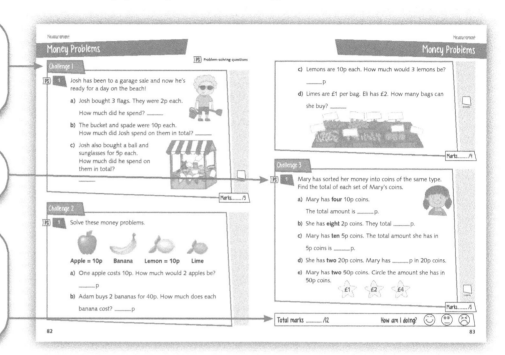

Starter test recaps skills covered in P1/P2.

Four progress tests throughout the book, allowing children to revisit the topics and test how well they have remembered the information.

Progress charts to record results and identify which areas need further revision and practice.

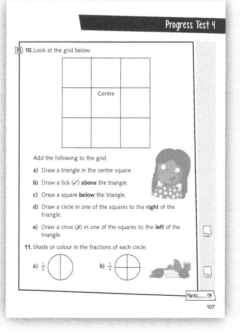

Answers for all the questions are included in a pull-out answer section at the back of the book.

Contents

Contents

ACKNOWLEDGEMENTS

The author and publisher are grateful to the copyright holders for permission to use quoted materials and images.

All illustrations and images are © Shutterstock and © HarperCollinsPublishers

Every effort has been made to trace copyright holders and obtain their permission for the use of copyright material. The author and publisher will gladly receive information enabling them to rectify any error or omission in subsequent editions. All facts are correct at time of going to press.

Published by Leckie
An imprint of HarperCollinsPublishers, Westerhill Road, Bishopbriggs, Glasgow G64 2QT

HarperCollins Publishers
Macken House, 39/40 Mayor Street Upper, Dublin 1, D01 C9W8, Ireland

© HarperCollinsPublishers Limited 2017

ISBN 978-0-00-866586-9

First published 2017

10 9 8 7 6 5 4 3 2 1

All rights reserved. No part of this publication may be reproduced, stored in a retrieval system, or transmitted, in any form or by any means, electronic, mechanical, photocopying, recording or otherwise, without the prior permission of Leckie.

Note for teachers: this book is also available as a downloadable pdf for unlimited school use: ISBN 978-0-00-826384-3

British Library Cataloguing in Publication Data. A CIP record of this book is available from the British Library.

Series Concept and Development: Michelle I'Anson Commissioning Editor: Richard Toms
Series Editor: Charlotte Christensen
Author: Alan Dobbs
Project Manager and Editorial: David Mantovani
Cover Design: Sarah Duxbury
Cover Illustration: Louise Forshaw
Inside Concept Design: Ian Wrigley
Text Design and Layout: Contentra Technologies Artwork: Collins and Contentra Technologies Production: Natalia Rebow

Printed in the UK, by Ashford Colour Press Ltd

1. Fill in the missing numbers. Check which way each set is counting and the steps it counts in.

a)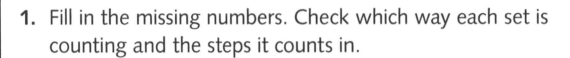

b)

c)

3 marks

2. What would 1 more and 1 less of each number be? Fill in the boxes.

Example:	1 less = 2	3	1 more = 4

a) [ ] **4** [ ]

b) [ ] **9** [ ]

c) [ ] **6** [ ]

d) [ ] **8** [ ]

e) [ ] **5** [ ]

f) [ ] **7** [ ]

6 marks

3. Tick (✓) the shapes that have $\frac{1}{2}$ coloured.

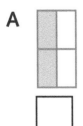 A   B   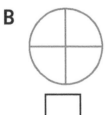 C   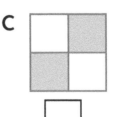 D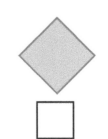

[ ]   [ ]   [ ]   [ ]

2 marks

4. Halve these numbers.

a)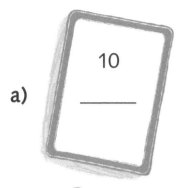
10
_____

b)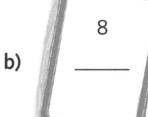
8
_____

c)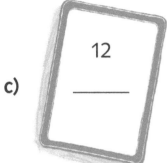
12
_____

d)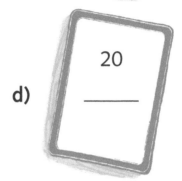
20
_____

4 marks

5. Are these sentences **true** or **false**?

a) When you count in twos, starting at 2, your answer is

always an even number. _____

b) If you add an odd number to an even number the answer

is always even. _____

c) 3 + 5 has the same total as 5 + 3. _____

3 marks

6. Write these words as numbers.

Example: Nine ___9___

a) Seven _____   **b)** Fifteen _____   **c)** Four _____

d) Twenty _____   **e)** Seventeen _____   **f)** Zero _____

6 marks

PS **7.** **a)** Sadiq had a hen and it laid 2 eggs every day. How many eggs did Sadiq's hen lay in 3 days? _____ eggs

b) Show how you worked this out.

_____

2 marks

8. Write these numbers as words.

Example: 13 ___Thirteen___

a) 12 _____

b) 10 _____

c) 7 _____

d) 6 _____

e) 0 _____

f) 18 _____

g) 11 _____

h) 5 _____

i) 9 _____

j) 16 _____

10 marks

PS **9.** Nell the pony eats five apples every day.

a) How many apples would she eat in 2 days? _____ apples

b) How did you work this out?

_____

c) How many apples would she eat in 3 days? _____ apples

d) How did you work this out?

_____

4 marks

10. Circle the calculation with an error in each set.

a)
```
1 + 1 = 2
2 + 1 = 3
3 + 1 = 5
4 + 1 = 5
```

b)
```
2 + 2 = 6
2 + 4 = 6
2 + 6 = 8
2 + 7 = 9
```

c)
```
10 – 5 = 5
12 – 5 = 7
14 – 5 = 10
15 – 5 = 10
```

3 marks

11. Draw hands on the clocks to show these times.

a) 3 o'clock

b) 9 o'clock

c) 5 o'clock

3 marks

12. Look at these splendid snakes.

A   B   C   D

 = 8 cm = 15 cm = 7 cm = 18 cm

a) Which snake is the shortest? _____

1 mark

b) Is snake A longer or shorter than snake B? _____

1 mark

c) Which snake is the longest? _____

1 mark

d) Put the snakes in order from shortest to longest.

shortest 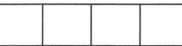 longest

2 marks

7

PS **13.** Mia is thinking of a number.
Her number is **half** of 20.

a) What is Mia's number? _____

b) How did you work this out?

_____

2 marks

14. Write the answer to each problem.

a) 8 sets of 2 = _____　　　**b)** 4 sets of 5 = _____

c) 3 sets of 10 = _____　　　**d)** 12 sets of 2 = _____

e) 5 sets of 5 = _____　　　**f)** 12 sets of 5 = _____

g) 5 sets of 10 = _____　　　**h)** 9 sets of 2 = _____

i) 4 sets of 10 = _____　　　**j)** 6 sets of 5 = _____

10 marks

PS **15.** Look at the busy bees below.

a) Count the busy bees. How many are there? _____

b) If two bees flew away, how many would remain? _____

c) If three bees joined the original group, how many would

there be now? _____

3 marks

16.a) Work out if these sequences are **correct** or **incorrect**.
Tick (✓) the correct sequences.

A  ☐

B  ☐

C  ☐

D  ☐

2 marks

b) Describe how these sequences are counting.

A	10	9	8	7	6	5

_____

B	2	4	6	8	10

_____

2 marks

9

c) Tick (✓) the sequence that is the odd one out.

A	5	10	15	20	25	30

☐

B	10	15	20	25	30	35

☐

C	12	14	16	18	20	22

☐

1 mark

d) How do you know this?

_____

_____

1 mark

17. Split these numbers to show tens and ones:

a) 15

Tens	Ones

b) 16

Tens	Ones

c) 29

Tens	Ones

3 marks

18. Add the following numbers.

a) 3 + 3 = _____

b) 4 + 4 = _____

c) 5 + 5 = _____

d) 6 + 6 = _____

4 marks

19. Add the flowers.

a)  +  = _____

b) 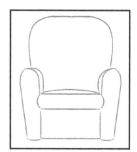 +  = _____

2 marks

20. Add 10 to each of these one-digit numbers.

a) 7 _____   **b)** 5 _____   **c)** 1 _____

3 marks

21. a) Draw lines to join each item to its estimated weight.

| about 3 kg | about 100 g | more than 10 kg | less than 20 g |

4 marks

b) Which item weighs the least? _____

1 mark

Marks.........../89

Numbers and Counting

Challenge 1

1 **a)** Tick the number that has the lowest value.

(15) (11) (13) (12) (14)

1 mark

b) Put these numbers in order from lowest to greatest value.

6   **3**   **11**   **2**

lowest | | | | | greatest

2 marks

2 **a)** Tick the number that has the greatest value.

| 14 | 24 | 22 | 32 | 18 |

1 mark

b) Put the numbers from part **a)** in order from least to greatest.

least | | | | | | greatest

3 marks

Marks............/7

Challenge 2

1 **a)** What is the value of the 1 in 19? _____

b) What value does the 3 have in 23? _____

c) What is the value of the 2 in 25? _____

d) What value does the 4 have in 43? _____

4 marks

2 Split these numbers into tens and ones.

a) 26

Tens	Ones

b) 26 = _____ + _____

2 marks

Numbers and Counting

3 Fill in the missing numbers.

	4			7	8		10		

13	14				18			21	22

2 marks

Marks.......... /8

Challenge 3

1 Fill in the missing numbers on this 100 square.

1		3				7	8		
11	12			15					20
		23	24					29	
		33			36				40
41							48		
				55				59	
		63				67			
						77	78		
			84						90
						97			

10 marks

Marks........ /10

Total marks .... /25   How am I doing?

13

Counting Forwards and Backwards

PS Problem-solving questions

Challenge 1

Use this number grid to help you count.

1	2	3	4	5	6	7	8	9	10
11	12	13	14	15	16	17	18	19	20

1 Start at **17** and count **back** the given amounts.

a) Count back 3 = _____   b) Count back 8 = _____

c) Count back 10 = _____   d) Count back 12 = _____

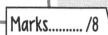

4 marks

2 Start at **3** and count **forwards** the given amounts.

a) Count forwards 4 = _____   b) Count forwards 7 = _____

c) Count forwards 9 = _____   d) Count forwards 12 = _____

4 marks

Marks.........../8

Challenge 2

PS **1** 10 birds are sitting in a tree.

3 of the birds fly away to look for food.
How many birds are still in the tree? _____

1 mark

Counting Forwards and Backwards

PS **2** Mia collects stickers. She has 12 in her collection. Mia gets 5 more stickers. How many does she have now? _____

1 mark

3 Complete the missing numbers on the number snake.

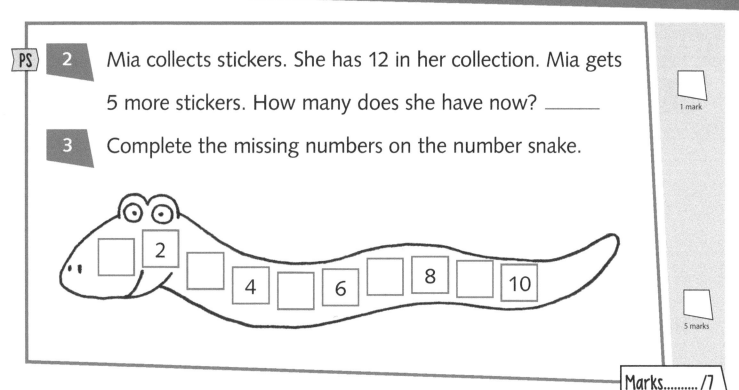

5 marks

Marks.......... /7

Challenge 3

1 Put these numbers in order from least to greatest.

| 15 | 14 | 4 | 19 | 5 | 20 | 13 |

least | | | | | | | | greatest

5 marks

2 Fill in the empty squares to complete these number grids. The first line is done for you.

a)

1	2	3
	12	

b)

23	24	25
		35

10 marks

Marks......... /15

Total marks .... /30   How am I doing?

Counting in Steps of 2, 5 and 10

PS Problem-solving questions

Challenge 1

1 Continue the numbers counting in steps of 2. Be careful, some are counting backwards!

a)
2 4 6 8

b)
15 13 11

2 marks

2 What steps are these numbers counting in?

a)

5	7	9	11

The numbers are counting in steps of _____.

b)

5	10	15	20

The numbers are counting steps of _____.

2 marks

Marks.......... /4

Challenge 2

1 Fill in the blank stepping stones. Make sure you count in **steps of 5**!

a)
15 30 40

b)
20 35

2 marks

Counting in Steps of 2, 5 and 10

2 These numbers are counting in steps of 5, but are they counting **forwards** or **backwards**?

a) | 20 | 15 | 10 | 5 | 0 |

The numbers are counting _____ .

b) | 14 | 19 | 24 | 29 |

The numbers are counting _____ .

2 marks

Marks.......... /4

Challenge 3

PS **1** Fill in the two missing numbers to complete each scarf. You will have to decide what steps they are counting in before you answer.

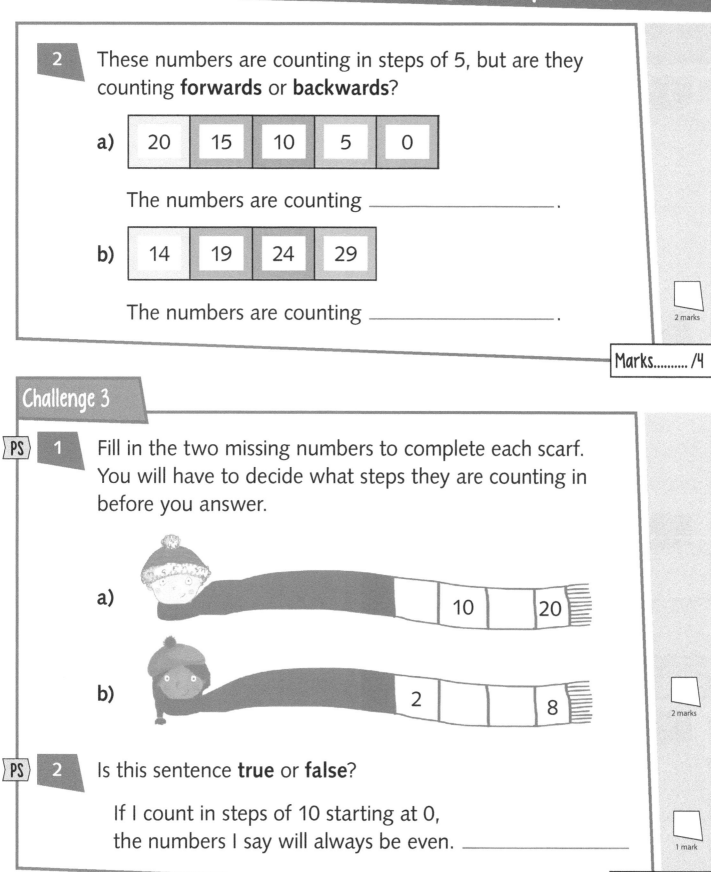

a) | | 10 | | 20 |

b) | 2 | | | 8 |

2 marks

PS **2** Is this sentence **true** or **false**?

If I count in steps of 10 starting at 0, the numbers I say will always be even. _____

1 mark

Marks.......... /3

Total marks .... /11   How am I doing?

Counting in Steps of 2, 5 and 10

 PS Problem-solving questions

Challenge 1

PS **1** Each pencil case holds 10 pencils. Count the total amount of pencils in lots of 10.

There are _____ pencils in total.

1 mark

2 How many lots of 10 are there in 20? _____

1 mark

3 Complete the grid. Make sure you are counting in lots of 10!

	20		
50			80

2 marks

4 Start at 10 and count **forwards** 3 lots of 10.

Which number are you at? _____

1 mark

Marks............/5

Challenge 2

1 Put the numbers in order to make a sequence counting **forwards** in **10s**.

30   50   10   20   40

least | | | | | | greatest

3 marks

Counting in Steps of 2, 5 and 10

2 Put the numbers in the correct order so they count **forwards** in **10s**.

least | | | | | greatest

3 marks

Marks.......... /6

Challenge 3

PS **1** Look at these buses.

a) Each bus has **10** seats. How many **lots of 10** seats are there? _____

b) How many **lots of 5** is that? _____

c) How many **sets of 2** seats are there on **1 bus**?

_____

d) How many **buses** would you need to have **30 seats**?

_____

4 marks

Marks.......... /4

Total marks .... /15   How am I doing?

Counting More and Less

PS Problem-solving questions

Challenge 1

1 Find 1 more and 1 less of these numbers.

a)	15	
b)	18	
c)	9	
d)	19	
e)	17	

1 less   **1 more**

2 What is 5 more than each number?

a) 2 _____   b) 4 _____   c) 11 _____

5 marks

3 marks

Marks.......... /8

Challenge 2

1 Which of these numbers has a greater value?

a) 3 or 7? _____   b) 17 or 12? _____

c) 25 or 19? _____   d) 10 or 15? _____

4 marks

2 Which of these numbers has a lower value?

a) 26 or 16? _____   b) 29 or 38? _____

c) 11 or 16? _____   d) 24 or 27? _____

4 marks

3 Write the number that is 10 more than 9. _____

1 mark

Marks.......... /9

Counting More and Less

Challenge 3

1 What would 10 less than each number be?

a) 14 _____

b) 26 _____

c) 40 _____

d) 32 _____

4 marks

PS **2** Answer these number problems.

a) Alan has **6 gold stars** in his book. Liz has **4 more** than

Alan. How many stars does Liz have? _____

b) Jacob has **2 apple trees**. He collects **15 apples** from the first tree. But the second tree has **5 fewer** than the first.

How many apples does the **second tree** have? _____

2 marks

PS **3** Steven takes **20 cupcakes** to school for the summer fair. Anna brings **10 more** than Steven.

a) How many cupcakes does Anna bring? _____

1 mark

b) How many cupcakes do they both bring in total? _____

Show how you worked this out. _____

2 marks

Marks.......... /9

Total marks .... /26

How am I doing?

Place Value

Challenge 1

1 Write the number that is shown on each abacus.

a) 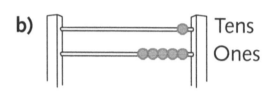 Tens Ones

_____

b) Tens Ones

_____

c) 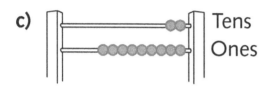 Tens Ones

_____

d) Tens Ones

_____

4 marks

2 Draw beads on each abacus to show these numbers.

a) 12 Tens Ones **b)** 24 Tens Ones

c) 35 Tens Ones **d)** 42 Tens Ones

4 marks

Marks.......... /8

Challenge 2

1 Partition these two-digit numbers into tens and ones.

Example: 32 = 30 + 2

a) 12 = _____ + _____   **b)** 25 = _____ + _____

c) 38 = _____ + _____   **d)** 46 = _____ + _____

e) 53 = _____ + _____

5 marks

Place Value

2 Write the two-digit numbers made with these tens and ones.

Example: 50 + 8 = <u>58</u>

a) 20 + 5 = _____

b) 30 + 1 = _____

c) 40 + 2 = _____

d) 30 + 7 = _____

 4 marks

Marks.........../9

Challenge 3

PS **1** This number is between 11 and 15.
What could the number be? _____

1 mark

PS **2** There are between 15 and 20 biscuits in this box.

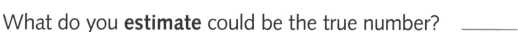 BISCUITS

What do you **estimate** could be the true number? _____

1 mark

PS **3** How many tens and ones do these two-digit numbers have?

a) 14: _____ ten, _____ ones

b) 24: _____ tens, _____ ones

c) 36: _____ tens, _____ ones

d) 48: _____ tens, _____ ones

4 marks

4 Put these numbers in order from least value to most.

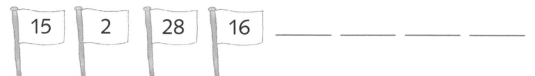 15 2 28 16 _____ _____ _____ _____

2 marks

Marks.........../8

Total marks ..../25   How am I doing?

Less Than, Greater Than and Equal To

Challenge 1

1 Sort the numbers into pairs so that they match the descriptions.

(17) (14) (20) (1) (20) (26) (13) (1) (12)

a) _____ is less than _____

b) _____ is greater than _____

c) _____ is equal to _____

3 marks

2 Choose your own numbers that match these descriptions:

a) _____ is greater than _____

b) _____ is less than _____

c) _____ is equal to _____

3 marks

Marks.......... /6

Challenge 2

1 Use the words in the boxes to compare the numbers.

| is less than | is greater than | is equal to |

Example: 6 <u>is greater than</u> 3

a) 15 _____ 18

b) 23 _____ 23

c) 20 _____ 14

3 marks

Less Than, Greater Than and Equal To

2 Use the words in the boxes to compare the numbers.

| is less than | is greater than | is equal to |

Example: Thirty-five <u>is less than</u> fifty-one

a) Twenty-seven _____ thirty-two

b) Twenty-one _____ twenty

c) Twenty-three _____ twenty-three

3 marks

Marks.......... /6

Challenge 3

PS **1** Choose the correct word from the boxes to make each sentence true.

| fewer | equal | more |

a) If there are 10 pencils and 10 pens they are

_____ .

b) Albi has 5 badges and Ben has 6 badges. Albi has

_____ badges than Ben.

c) 10 fish meet 12 tadpoles. There are _____ tadpoles than fish.

d) Fran collects 20 blueberries and 15 redcurrants. She has

_____ redcurrants than blueberries.

4 marks

Marks.......... /4

Total marks .... /16   How am I doing? 😊 😐 😣

Doubling, Halving, Adding and Subtracting

Challenge 1

1 **Double** and **halve** these numbers using each set of three cards to help you.

 **Example:** 2, 4, 2   Double <u>2</u> is <u>4</u>, half of <u>4</u> = <u>2</u>

a) 5 10 5   **b)** 7 14 7

a) Double _____ is _____, half of _____ = _____

b) Double _____ is _____, half of _____ = _____

2 marks

2 Now try these numbers without any help!

 **Example:** 3: Double 3 = <u>3</u> + <u>3</u> = <u>6</u>

a) 2: Double 2 = _____ + _____ = _____

b) 6: Double 6 = _____ + _____ = _____

c) 10: Double 10 = _____ + _____ = _____

d) 7: Double 7 = _____ + _____ = _____

4 marks

Marks.......... /6

Challenge 2

1 Complete the fact families for the sets of numbers given on the next page.

Example: 2, 3 and 5 give the facts <u>2</u> + <u>3</u> = <u>5</u>,
<u>3</u> + <u>2</u> = <u>5</u>, <u>5</u> − <u>3</u> = <u>2</u> and <u>5</u> − <u>2</u> = <u>3</u>

Doubling, Halving, Adding and Subtracting

a) 5 10 15   b) 3 7 10

c) 2 3 5   d) 20 10 30

a) _____ + _____ = _____   _____ + _____ = _____

_____ – _____ = _____   _____ – _____ = _____

b) _____ + _____ = _____   _____ + _____ = _____

_____ – _____ = _____   _____ – _____ = _____

c) _____ + _____ = _____   _____ + _____ = _____

_____ – _____ = _____   _____ – _____ = _____

d) _____ + _____ = _____   _____ + _____ = _____

_____ – _____ = _____   _____ – _____ = _____

4 marks

Marks.......... /4

1 Solve these addition number sentences.

a) $8 = \underline{\hspace{1cm}} + 2$ b) $\underline{\hspace{1cm}} + 4 = 10$

c) $7 = 4 + \underline{\hspace{1cm}}$ d) $15 = \underline{\hspace{1cm}} + 9$

4 marks

2 Solve these subtraction number sentences.

a) $15 - 3 = \underline{\hspace{1cm}}$ b) $19 - 6 = \underline{\hspace{1cm}}$

c) $12 - 7 = \underline{\hspace{1cm}}$ d) $13 - 2 = \underline{\hspace{1cm}}$

4 marks

Marks.......... /8

Total marks .... /18   How am I doing?

Solving Number Problems

Challenge 1

Answer these addition and subtraction problems. You could draw pictures to help solve them.

 1 Matt has 8 stickers. He gives 6 stickers to his friends. How many stickers does Matt have left?

_____ – _____ = _____

1 mark

 2 Sita has 10 colourful beads. Her sister gives her 7 more beads. How many beads does Sita have altogether?

_____ + _____ = _____

1 mark

 3 Amy has 12 caterpillars in her butterfly house. 3 of them change into butterflies. How many caterpillars remain?

_____ – _____ = _____

1 mark

Marks.........../3

Challenge 2

 1 **a)** Rose has 3 boxes of apples. Each box holds 10 apples.

How many apples does Rose have in total?

_____

b) Show this as an addition sum.

_____ + _____ + _____ = _____

2 marks

Solving Number Problems

PS **2** Jan collects 12 conkers. Danny finds 8 conkers.

    a) How many more conkers does Jan have? _____

    b) Show how you worked out the difference. _____

2 marks

PS **3** There are 9 girls in class 4 and 7 girls in class 5.

    How many girls are there in total? _____

1 mark

Marks.......... /5

Challenge 3

1 Write the totals of these numbers. Check your answers by adding each set of numbers in a different order.

    a) 5 + 3 + 1

    _____ + _____ + _____ = _____

    _____ + _____ + _____ = _____

    b) 3 + 4 + 2

    _____ + _____ + _____ = _____

    _____ + _____ + _____ = _____

    c) 6 + 1 + 5

    _____ + _____ + _____ = _____

    _____ + _____ + _____ = _____

3 marks

Marks.......... /3

Total marks .... /11   How am I doing?

Using Two-Digit Numbers

PS Problem-solving questions

Challenge 1

1 Look at the buns. There are 10 in total. Write eleven addition number facts that total 10.

Example: 9 + 1 = 10

_____ + _____ = 10	
_____ + _____ = 10	_____ + _____ = 10
_____ + _____ = 10	_____ + _____ = 10
_____ + _____ = 10	_____ + _____ = 10
_____ + _____ = 10	_____ + _____ = 10
_____ + _____ = 10	_____ + _____ = 10

11 marks

Marks.......... /11

Challenge 2

1 Create ten different two-digit numbers using the one-digit numbers on the cards.

1   2   5   6   8   3   7   4   9

a) _____   b) _____   c) _____   d) _____

e) _____   f) _____   g) _____   h) _____

i) _____   j) _____

10 marks

30

Using Two-Digit Numbers

2 Add 10 to each number to change them into two-digit numbers.

a) 3 _____   **b)** 4 _____   **c)** 6 _____

d) 1 _____   **e)** 8 _____

5 marks

Marks......... /15

Challenge 3

PS **1** There are 10 flowers left. Write five subtraction number facts to show how many flowers there could have been to begin with.

Example: 16 – 6 = 10

_____ – _____ = 10	_____ – _____ = 10
_____ – _____ = 10	_____ – _____ = 10
_____ – _____ = 10	

5 marks

PS **2** These leaves are numbered using two-digit numbers. Write the correct numbers on the blank leaves.

4 marks

Marks.......... /9

Total marks .... /35   How am I doing?

Solving Missing Number Problems

1 Use the number grid to complete these additions and subtractions by writing in the missing numbers.

1	2	3	4	5	6	7	8	9	10	11	12
13	14	15	16	17	18	19	20	21	22	23	24

a) 4 + _____ = 10

b) 24 – _____ = 20

c) _____ + 10 = 12

d) 19 + _____ = 20

e) _____ + 16 = 20

f) 15 – _____ = 12

g) 20 – 8 = _____

h) 13 – _____ = 3

i) _____ + 5 = 15

9 marks

Marks.......... /9

1 Complete this number grid by filling in the missing numbers.

2		4		
	8			11
		14		
17			20	
	23			26

5 marks

Solving Missing Number Problems

2 Fill in the missing number operation symbols (**–** or **+**).

a) 3 _____ 7 = 10

b) 5 _____ 10 = 15

c) 15 _____ 10 = 5

d) 17 = 10 _____ 7

e) 17 = 20 _____ 3

5 marks

Marks.........../10

Challenge 3

1 Casey is selling bows at a garage sale, but she has lost her price tags! She knows that purple bows are 5p. Can you help Casey to remember the price of her other bows?

a) Pink bows are 3p more than purple bows.

Pink bows cost _____p.

b) Blue bows are 10p more than purple bows.

Blue bows cost _____p.

c) Casey's green bows are 2p more than blue bows.

Green bows cost _____p.

d) Yellow bows cost the same as a pink bow and a blue bow added together.

A yellow bow should cost _____p.

e) An orange bow should cost 1p less than a purple bow.

An orange bow is _____p.

5 marks

Marks.........../5

Total marks ..../24

How am I doing?

Mixed Number Problems

 PS Problem-solving questions

1 True or false? Write **true** or **false** next to these additions.

a) 12 + 6 = 20 _____

b) 14 + 6 = 20 _____

c) 4 + 12 = 15 _____

3 marks

2 Spot the mistake! Write the correct answer for each addition.

a) 12 + 7 = 20 ✗   12 + 7 = _____

b) 15 + 6 = 19 ✗   15 + 6 = _____

c) 3 + 12 = 16 ✗   3 + 12 = _____

3 marks

Marks.......... /6

PS **1**
a) Hamid is thinking of a number.
The number is 10 less than 25.
What is Hamid's number?   _____

b) Jane is thinking of a number between
14 and 21. Write the possible numbers
that Jane could be thinking of.

_____

c) Dan and Helen both have a number. Helen's
number is 10 more than Dan's. Dan's number
is 12. What is Helen's number?   _____

d) Adam has 10 stickers. He gives 5 stickers away.
How many stickers does Adam have left?   _____

4 marks

Marks.......... /4

34

Mixed Number Problems

Challenge 3

1 Put a tick (✓) next to the correct subtractions, and a cross (✗) next to the ones that are wrong.

a) 13 – 7 = 6 ☐

b) 12 – 5 = 6 ☐

c) 20 – 12 = 8 ☐

3 marks

PS **2** a) Chandra has 12 cherries. He gives 5 to Simon. How many cherries does Chandra have left? _____

b) Write the subtraction that shows this.

_____ – _____ = _____

2 marks

PS **3** Write **true** or **false** next to these statements.

a) If you subtract a smaller number from a greater number the total is always less. _____

b) Subtracting an odd number from an even number always gives you an even number. _____

2 marks

PS **4** a) Harry subtracted 10 from his starting number. His answer was 10. What was Harry's starting number? _____

1 mark

b) Kira's answer is 5. Write four subtractions that she could have made.

_____ – _____ = 5   _____ – _____ = 5

_____ – _____ = 5   _____ – _____ = 5

4 marks

Marks......... /12

Total marks .... /22   How am I doing?   😊   😐   😣

35

1. Double these numbers.

a) 5 _____   **b)** 4 _____

c) 10 _____   **d)** 9 _____

4 marks

2. What is half of each number?

a) 2 _____   **b)** 4 _____

c) 10 _____   **d)** 20 _____

4 marks

3. Put any two of these one-digit numbers together to form eight different two-digit numbers.

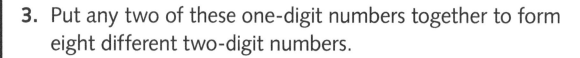

Example: Digits **1** and **3** make the two-digit number **13**

_____ _____ _____ _____ _____ _____ _____ _____

5 marks

PS 4. a) Rob counts 3 butterflies in his garden. He sees 4 more. How many butterflies can Rob now see?

_____

b) Three of the butterflies fly away. How many

butterflies can Rob see now? _____

c) Six new butterflies appear. What is the total number of

butterflies now? _____

3 marks

5. Swap the order of the numbers to make equal addition facts.

a) 9 + 3 = 12 _____

b) 2 + 8 = 10 _____

c) 7 + 4 = 11 _____

d) 6 + 10 = 16 _____

4 marks

6. Circle the card that has the highest value.

19   10   13   20   14

1 mark

7. Write these numbers using digits.

a) Seven _____   **b)** Fourteen _____

c) Sixteen _____   **d)** Twenty _____

4 marks

8. Tom has some large marbles and small marbles. The large marbles each have a value of 10 and the small marbles are 1s. Calculate the numbers that Tom has made using the marbles.

a)  _____

b)  _____

c)  _____

3 marks

9. What is 5 less than each number?

a) 9 _____   **b)** 10 _____

c) 20 _____   **d)** 16 _____

4 marks

10. Write these numbers as words.

    a) 5 _____

    b) 7 _____

    c) 9 _____

    d) 2 _____

    e) 12 _____

5 marks

11. Complete this addition grid. Make sure that each **column** and each **row** has a total of 10. You could use objects to help you.

1	5	
1		6
	2	0

3 marks

12. Look at the cards below.

    1 2 5 6 8 3 7 4 9

    a) Using any two numbers from the cards, what is the lowest value, two-digit number possible? _____

    b) Using any two numbers from the cards, what is the highest value two-digit number possible? _____

    c) Make an **odd** two-digit number from the numbers on the cards. _____

    d) Make an **even** two-digit number from the numbers on the cards. _____

4 marks

13. Find the fact families for these sets of numbers.

> **Example:** 13, 7, 20 gives <u>13</u> + <u>7</u> = <u>20</u>, <u>7</u> + <u>13</u> = <u>20</u>,
> <u>20</u> − <u>7</u> = <u>13</u>, <u>20</u> − <u>13</u> = <u>7</u>

a) 17, 3, 20	**b)** 14, 5, 19	**c)** 11, 4, 15

a) _____ + _____ = _____   _____ + _____ = _____

   _____ − _____ = _____   _____ − _____ = _____

b) _____ + _____ = _____   _____ + _____ = _____

   _____ − _____ = _____   _____ − _____ = _____

c) _____ + _____ = _____   _____ + _____ = _____

   _____ − _____ = _____   _____ − _____ = _____

3 marks

PS **14.** Answer these number problems.

a) Put the stars into groups of 5.

b) Shen has 7 toy cars. He wins 6 more. How many cars

   does Shen have now? _____

2 marks

Marks........ /49

What is Multiplication?

Challenge 1

1 Use the buttons to answer the questions.

a) Count the holes in lots of 2. There are _____ holes in total.

b) Write this as a multiplication.

_____ sets of _____ = _____

2 marks

2 Write the multiplication for each of these additions.

a) 2 + 2 + 2 = 6   _____ sets of _____ = _____

b) 5 + 5 = 10   _____ sets of _____ = _____

c) 10 + 10 + 10 = 30   _____ sets of _____ = _____

3 marks

Marks.......... /5

Challenge 2

1 There are two possible multiplications shown by this grouping of blocks. What are they?

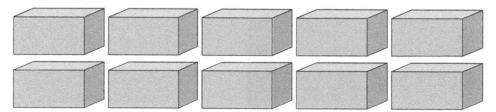

_____ sets of _____ = _____

_____ sets of _____ = _____

2 marks

What is Multiplication?

2 Draw blocks to show each of these multiplications.

a) 4 sets of 2	b) 3 sets of 5	c) 2 sets of 10

6 marks

Marks.......... /8

Challenge 3

1 Complete the table. The first row has been done for you.

Addition	Number of sets
2 + 2 + 2 + 2 + 2 = 10	5 sets of 2
	3 sets of 5
10 + 10 + 10 = 30	
	4 sets of 5
	5 sets of 10
2 + 2 + 2 + 2 + 2 + 2 = 12	

5 marks

Marks.......... /5

Total marks .... /18   How am I doing?

What is Division?

Problem-solving questions

Challenge 1

PS | **1** Answer these division problems. You could use objects to help you.

a) There are 4 fruit sticks to share between 2 people. Divide them so that each person has the same amount. How many fruit sticks would each person get?

Each person would get _____ fruit sticks.

b) If you share 6 pencils between two people, how many would each person get?

Each person would get _____ pencils.

c) If you share 10 stickers between two people, how many would each person get?

Each person would get _____ stickers.

d) How many biscuits would you need for 2 people to have 4 biscuits each?

You would need _____ biscuits in total.

4 marks

Marks.........../4

Challenge 2

PS | **1** Share these strawberries equally between 2 people.

Each person will get _____ strawberries.

1 mark

What is Division?

2 Here are 2 very hungry children!
Share 12 apple slices equally so each
child gets the same amount.

Each child will get _____ apple slices.

1 mark

PS **3** Share 9 cookies equally between 3 people.

Each person will get _____ cookies.

1 mark

Marks.........../3

Challenge 3

1 Complete this table by sharing the number of items
between the number of people. The first row has been
done for you.

Number of items	Number of people	They each get
6	2	3
14	2	
16	2	
18	2	
20	2	

4 marks

Marks.........../4

Total marks .... /11   How am I doing?

2, 5 and 10 Times Tables – Odds and Evens

1 Complete this grid showing 2, 5 and 10 times tables. One row has been done for you.

2 times 1 = 2	5 times 1 = 5	10 times 1 = 10
2 times 2 = 4		
	5 times 3 = 15	10 times 3 = 30
2 times 4 = 8		
		10 times 5 = 50
	5 times 6 = 30	

9 marks

Marks......... /9

1 Colour in the odd numbers and draw stripes through the even numbers. One column has been done for you.

1	2	3	4	5
6	7	8	9	10
11	12	13	14	15
16	17	18	19	20

4 marks

2, 5 and 10 Times Tables – Odds and Evens

2 Is each sentence **true** or **false**?

a) When you count in lots of 2, starting from 2, the numbers you say are sometimes **odd numbers**.

_____

b) When you count in lots of 2, starting from 2, the numbers you say are always even numbers.

_____

2 marks

Marks.......... /6

Challenge 3

Multiplication is **commutative**. It has the same **product** if the order of the **numbers** is changed.

Example: 3 times 2 = 6 and 2 times 3 = 6

1 Write the **commutative** of each multiplication.

a) 4 times 2 = 8 _____

b) 1 times 10 = 10 _____

2 marks

2 Each **array** shows two **multiplications**. Write the two multiplications shown by each array.

a)

b)

_____   _____

_____   _____

4 marks

Marks.......... /6

Total marks .... /21   How am I doing?

Division Problems

PS Problem-solving questions

Challenge 1

Division is not **commutative**. It **cannot** be done in any order.

PS **1** Look at the strawberries.

There are 6 strawberries split into 2 groups of 3. Write down the division that you can see.

_____ shared by _____ = _____

1 mark

2 Use each set of numbers to write a correct division.

a) 10 2 5   _____ shared by _____ = _____

b) 20 10 2   _____ shared by _____ = _____

c) 16 2 8   _____ shared by _____ = _____

3 marks

3 Rearrange each set of numbers to make a correct division. Be careful, they are tricky!

a) 10 20 2   _____ shared by _____ = _____

b) 2 4 8   _____ shared by _____ = _____

c) 12 6 2   _____ shared by _____ = _____

3 marks

Marks............/7

46

Division Problems

Challenge 2

1 Tao has grown 20 sunflowers. He wants to share them with his friends.

a) If Tao shares his sunflowers between

2 friends, how many will they get each? _____

b) Write this as a division. _____ shared by _____ = _____

c) If 4 friends share the sunflowers, they would get _____ each.

d) Write this as a division. _____ shared by _____ = _____

4 marks

Marks.......... /4

Challenge 3

1 The coins below show three **different** ways that Jenny could have **20p**. Look at the coins and then complete the sentences.

a) 20p is equal to _____ 10p coins.

b) 20p is equal to _____ 5p coins.

c)

20p is equal to _____ 2p coins.

3 marks

Marks.......... /3

Total marks .... /14

How am I doing?

More Doubling, Halving and Dividing

1 Multiply these numbers by 2 to double them.

Example: Double 4 is <u>8</u>

a) Double 5 is _____

b) Double 10 is _____

c) Double 3 is _____

3 marks

2 Now divide these numbers by 2 to halve them.

Example: Half of 10 is <u>5</u>

a) Half of 4 is _____

b) Half of 8 is _____

c) Half of 10 is _____

3 marks

Marks.......... /6

1 Count these busy bees.

a) How many bees are there in total? _____

b) How many bees would half of this amount be? _____

2 marks

More Doubling, Halving and Dividing

2 Now count these flowers.

a) How many flowers are there in total? _____

b) How many flowers would there be if you doubled the group? _____

2 marks

Marks.......... /4

Challenge 3

1 Count these apples and then answer the questions that follow.

a) How many groups of 5 are there? _____

b) How many groups of 10 are there? _____

c) How many groups of 2 are there? _____

d) How many groups of 20 are there? _____

4 marks

Marks.......... /4

Total marks .... /14   How am I doing?

Solving Multiplication and Division Problems

PS Problem-solving questions

Challenge 1

PS **1** Allen had a hen and it laid 1 egg every day.

   a) How many eggs did Allen's hen lay in 5 days? _____

   b) How many eggs did Allen's hen lay in 10 days?

   _____

2 marks

PS **2** Henry had 2 pizzas. He cut them into quarters.

   a) How many slices of pizza did Henry have in total? _____

   b) Henry ate half of the full amount.
   How many slices of pizza did he eat? _____

2 marks

PS **3** Ravi can blow up 3 balloons in one minute.

   a) How many balloons can he blow up in 2 minutes? _____

   b) How many can he blow up in 5 minutes? _____

2 marks

Marks.......... /6

Challenge 2

PS **1** This is sports club!

   a) Count how many children are in the club.

   There are _____ children.

Solving Multiplication and Division Problems

b) Each child runs with a partner.
How many pairs of two can you count? _____

c) How many lots of 5 is this? _____

Pens

PS 2 Amir had 5 pens. Libby had double the amount.

a) How many pens did Libby have? _____

b) How many lots of 5 did Amir and Libby have in total?

_____

3 marks

2 marks

Marks.......... /5

Challenge 3

PS 1 Complete the table by answering the problems. The first one has been done for you.

Number problem	Answer
One person has 2 legs. How many legs do 4 people have?	8 legs
a) Two ladybirds have 8 spots. How many spots does 1 ladybird have?	
b) There are 10 boys in class 5. Class 6 has double the amount of boys. How many boys are in class 6?	
c) Sati gets 10 stickers each week. How many stickers will she get in 3 weeks?	

3 marks

Marks.......... /3

Total marks .... /14   How am I doing?

Numbers All Around Us

Challenge 1

PS **1** Beth and Judy live on this street.

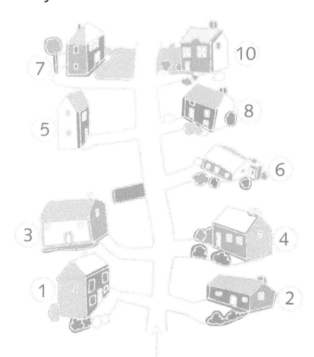

Start here

a) Beth lives at number 4. Judy lives opposite Beth.

What is the number of Judy's house? _____

b) Their friend Freddie lives at the third house on the right.

What is the number of Freddie's house? _____

c) Judy's gran lives at number 7.

Which house is opposite her gran's house? _____

d) Do the houses on the right have odd or even numbers?

_____

e) Do the houses on the left have odd or even numbers?

_____

5 marks

Marks.......... /5

52

Numbers All Around Us

Challenge 2

1 There are 10 children on a school bus.

a) 6 children get off at the first stop.
How many children are still on the bus? _____

b) More children get on and the bus
has 10 again. Of the 10 children, 5 are
girls. How many boys are on the bus? _____

c) At the next stop, 3 children get off the bus.
How many children are left? _____

3 marks

Marks.........../3

Challenge 3

 PS

1 Seta has 20 sweets in a bag.

a) 5 of the sweets are red. How many are **not** red? _____

Show how you worked this out as a subtraction.

_____ – _____ = _____

2 marks

b) 10 sweets are yellow. The rest of the sweets are green.
How many sweets are green?
(Don't forget the red sweets!) _____

1 mark

c) Seta shares the sweets with her friend.
How many sweets do they get each? _____

1 mark

Marks........../4

Total marks ..../12   How am I doing?

More Mixed Number Problems

PS Problem-solving questions

Challenge 1

1 How many groups of 2 are there in the sets of cars?

a)  = _____ groups of 2.

b) = _____ groups of 2.

c) = _____ groups of 2.

3 marks

Marks............/3

Challenge 2

PS **1** Share the pencils on the right equally between

2 people. Each person would get _____ pencils.

1 mark

2 Fill in the blanks to complete these sentences.

a) 8 shared by 2 equals _____.

b) 10 shared by 2 equals _____.

c) 12 shared by 2 equals _____.

3 marks

3 True or false? Write **true** or **false** next to these statements.

a) If you share 20 between 2 people, they
must each get different amounts. _____

More Mixed Number Problems

b) You can share an even number between 2 equally. _____

2 marks

Marks.......... /6

Challenge 3

1 Fill in the missing numbers.

a) 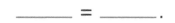 + + + = _____

b) The addition above can be written as _____ groups of

_____ = _____ .

2 marks

2 Now answer this one.

a) + = _____

b) The addition above can be written as _____ groups of

_____ = _____ .

2 marks

3 True or false? Tick (✓) the **true** multiplications and put a cross (✗) next to the ones that are **false**.

a) 2 times 5 = 10 ☐

b) 5 times 10 = 30 ☐

c) 5 times 2 = 12 ☐

d) 4 times 5 = 20 ☐

4 marks

Marks.......... /8

Total marks .... /17   How am I doing?

Halves as Fractions

Challenge 1

1 Shade in one half of each circle.

a)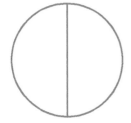

b)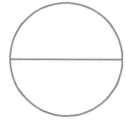

c)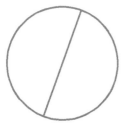

d)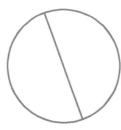

4 marks

2 How many halves are there in 2 circles? _____

1 mark

Marks.......... /5

Challenge 2

1 Shade one half of each square.

a)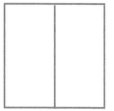

b)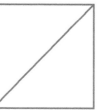

c)

d)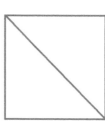

4 marks

2 How many halves are there in 4 squares? _____

1 mark

Marks.......... /5

Halves as Fractions

1 For each shape, put a tick (✓) in the box if $\frac{1}{2}$ is shaded or a cross (✗) in the box if $\frac{1}{2}$ is not shaded.

a) 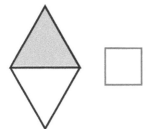 □

b) 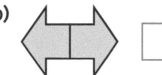 □

c) 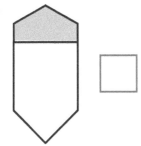 □

d) 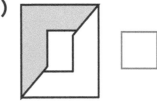 □

e) 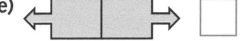 □

f) 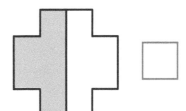 □

6 marks

Marks.......... /6

Total marks .... /16   How am I doing?

Quarters as Fractions

Challenge 1

1 Look at these shapes. They are divided into four equal parts. Colour $\frac{1}{4}$ of each shape.

a)   b)   c)  d)

4 marks

2 Now colour $\frac{1}{4}$ of each of these shapes.

a) 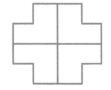 b)  c)  d)

4 marks

3 Tick (✓) the shapes that have $\frac{1}{4}$ shaded.

A   ☐   B 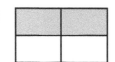 ☐   C 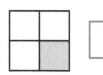 ☐

2 marks

Marks.........../10

Challenge 2

1 Each pizza is cut into quarters. How much of each pizza is left? Draw lines to match each fraction with the correct pizza.

a)   b)   c)   d)   e)

| 1 quarter | 2 quarters | 3 quarters | 0 quarters | 4 quarters |

5 marks

Quarters as Fractions

2 Look again at the pizzas. How many **quarters** of each pizza have been eaten?

a) _____ b) _____ c) _____ d) _____ e) _____

5 marks

Marks........ /10

Challenge 3

1 Tick (✓) the shapes that have $\frac{1}{4}$ shaded and put a cross (✗) next to the ones that have $\frac{1}{2}$ shaded.

a)

b)   c)

d)   e)

5 marks

Marks.......... /5

Total marks .... /25   How am I doing?

Fractions of Groups

1 Find the fractions of these groups.

A

B

C

D

a) Draw a circle around $\frac{1}{2}$ of group A. How many buttons have you circled? _____

b) Draw a circle around $\frac{1}{2}$ of group B. How many buttons have you circled? _____

c) Circle $\frac{1}{2}$ of group C. How many buttons have you circled? _____

d) $\frac{1}{2}$ of group D is _____ buttons.

e) How many buttons are in $\frac{1}{4}$ of group B? _____

5 marks

Marks.......... /5

Fractions of Groups

Challenge 2

1 Take a look at the group of shells and answer the questions.

a) Draw a red line around $\frac{1}{4}$ of the shells. How many shells have you drawn around? _____

b) Show half of the group using a blue line. How many shells is this? _____

c) How many sets of 4 shells can you make from the whole group? _____

3 marks

Marks.........../3

Challenge 3

1 Complete the table by working out the fractions.

Items	Amount in $\frac{1}{4}$	Amount in $\frac{1}{2}$
4 stickers		
20 buns		
8 straws		

6 marks

Marks........../6

Total marks ..../14   How am I doing?

Fractions of Numbers

Challenge 1

1 Write the numbers that make these sentences correct.

a) $\frac{1}{2}$ of 2 is _____

b) $\frac{1}{2}$ of 6 is _____

c) $\frac{1}{4}$ of 4 is _____

d) $\frac{1}{4}$ of 8 is _____

4 marks

2 Which is greater, $\frac{1}{2}$ or $\frac{1}{4}$? Draw lines to show which fraction is greater and which is smaller.

smallest   greatest

1 mark

Marks.......... /5

Challenge 2

1 Write the numbers that make these sentences correct.

a) $\frac{1}{2}$ of 12 is _____.

b) $\frac{1}{4}$ of 12 is _____.

c) $\frac{1}{2}$ of 16 is _____.

d) $\frac{1}{2}$ of 10 is _____.

e) $\frac{1}{4}$ of 20 is _____.

5 marks

Fractions of Numbers

2 Are these sentences **true** or **false**?

a) $\frac{1}{2}$ of 10 is 6. _____

b) $\frac{1}{4}$ of 20 is 5. _____

c) $\frac{1}{2}$ of 16 is 8. _____

d) $\frac{1}{4}$ of 16 is 3. _____

e) $\frac{1}{2}$ of 18 is 9. _____

5 marks

Marks......... /10

Challenge 3

1 Answer these fraction questions.

a) A quarter of this number is 2. What is the number?

_____

b) Half of this number is 10. What is the number?

_____

c) A quarter of the number equals 5. What is the whole

number? _____

d) $\frac{1}{4}$ of this number equals 4. What is the number?

_____

4 marks

Marks.......... /4

Total marks .... /19

How am I doing?

Fractions All Around Us

Challenge 1

PS **1** Oliver empties his money box. These are the coins he has.

a) How many 2p coins does Oliver have?

_____

b) How many 5p coins does he have?

_____

c) Oliver has _____ 1p coins.

3 marks

Marks........./3

Challenge 2

PS **1** This is Tomas's fish tank.

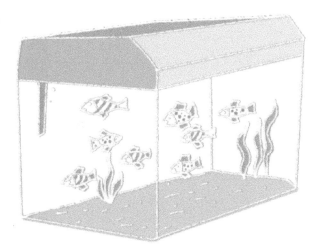

Fractions All Around Us

a) How many fish does Tomas have in total? _____

b) Half of the fish are striped. How many fish have

stripes? _____

c) Half of the fish have spots. How many fish are

spotted? _____

d) What would be a quarter of the total number of fish?

_____

4 marks

Marks.......... /4

Challenge 3

PS   1   Razz, Bell and Deli have had a snail race!
Bell's snail came first and travelled 12 cm in 12 minutes.

a) Razz's snail only went half the distance of Bell's.

Razz's snail travelled _____ cm.

b) Deli's snail went a quarter of 12 cm. How far did Deli's

snail go? _____ cm

c) Bell's snail travelled 1 cm every minute. How far had her

snail travelled after 2 minutes? _____ cm

d) How far had Bell's snail gone after 4 minutes? _____ cm

4 marks

Marks.......... /4

Total marks .... /11   How am I doing? 😊 😐 😣

1. Solve these addition problems.

   a) 3 + 6 = _____

   b) 10 + 5 = _____

   c) 7 + 4 = _____

   d) 9 + 2 = _____

   4 marks

2. Partition each two-digit number into tens and ones.

   a) 27 = _____ tens, _____ ones

   b) 19 = _____ tens, _____ ones

   c) 21 = _____ tens, _____ one

   d) 25 = _____ tens, _____ ones

   e) 15 = _____ tens, _____ ones

   5 marks

PS 3. Find the fractions.

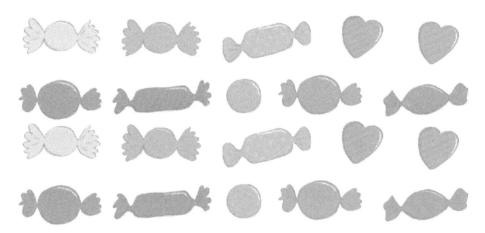

   a) How many would half of these sweets be?

   _____

   b) How many sweets would there be in $\frac{1}{4}$ of this group?

   _____

   c) How many sweets are there all together? _____

   3 marks

4. Find half of each number by dividing.

    a) Half of sixteen is _____ .

    b) Half of twenty is _____ .

    c) Half of 8 is _____ .

    d) Half of 10 is _____ .

    e) Half of 14 is _____ .

5 marks

5. Double each number by multiplying.

    a) Double 2 is _____ .

    b) Double 5 is _____ .

    c) Double 10 is _____ .

    d) Double 8 is _____ .

    e) Double 7 is _____ .

5 marks

6. Colour these **arrays** to show the following calculations.

    a) 4 times 2 = 8   **b)** 5 times 2 = 10

2 marks

PS **7.** Hopscotch the rabbit eats 2 carrots every day. How many carrots would he eat in:

    a) 2 days? _____

    b) 5 days? _____

    c) 10 days? (That's a lot of carrots!) _____

3 marks

8. Count up using lots of 5 to complete the flowerpots.

| 0 | 5 | | | | |

4 marks

9. Show 1 more and 1 less than each number.

a)

| 1 less = _____ | 13 | 1 more = _____ |

b)

| 1 less = _____ | 11 | 1 more = _____ |

c)

| 1 less = _____ | 19 | 1 more = _____ |

3 marks

10. Use the words in the boxes to make each number sentence correct.

| is less than | is greater than | is equal to |

a) 8 _____ 5

b) 7 _____ 17

c) 19 _____ 18

d) 16 _____ 16

e) 10 _____ 11

5 marks

11. Add the numbers. Check your answers by adding in a different order.

a) 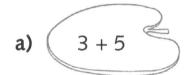 3 + 5

_____ + _____ = _____

_____ + _____ = _____

b) 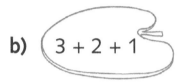 3 + 2 + 1

_____ + _____ + _____ = _____

_____ + _____ + _____ = _____

4 marks

PS **12.** Find the fractions of these groups.

A

B

C

a) How many apples are in $\frac{1}{4}$ of group A? ☐

b) How many apples are in $\frac{1}{4}$ of group B? ☐

c) How many apples are in $\frac{1}{2}$ of group C? ☐

d) How many apples are in $\frac{1}{2}$ of group A? ☐

4 marks

Marks........ /47

Measuring Length and Height

Challenge 1

1 These slippery snakes are different lengths.

A   B   C   D

 = 5 cm   = 10 cm    = 8 cm    = 12 cm

a) Which snake is the **longest**? _____

1 mark

b) Which snake is the **shortest**? _____

1 mark

c) Is snake C **longer** or **shorter** than snake D? _____

1 mark

d) Put the snakes into **letter** order from shortest to longest.

_____ _____ _____ _____

2 marks

e) How long would snakes **A** and **B** be together?

_____ cm

1 mark

PS **2** a) Take a look at Jill the giraffe. Write

down her height in metres. _____ m

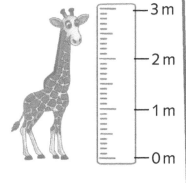

b) Her brother, Jack, is exactly
1 m smaller than Jill.

How tall is Jack? _____ m

c) Jill has a baby called Jem. Jem is the **same** height as

Jack. What is the height of Jem? _____ m

d) Jill's dad Ginger was a very tall giraffe.
He was 5 metres from head to hoof!

How much taller was Ginger than Jill? _____ m

4 marks

Marks.........../10

Measuring Length and Height

Challenge 2

1 Draw lines to match the items to their heights.

| about 1 m | less than 1 m | about 3 m | over 5 m | less than 1 m |

5 marks

2 Use a cm ruler to measure these shoelaces.

a)  = _____ cm

b)  = _____ cm

2 marks

Marks.......... /7

Challenge 3

1 Complete each number track.

a)
| 1 cm | 2 cm | 3 cm | | | 6 cm |

b)
| 5 cm | | | | 25 cm | |

c)
| 3 cm | | 7 cm | 9 cm | | |

3 marks

Marks.......... /3

Total marks .... /20   How am I doing?

71

Measuring Weight and Capacity

PS Problem-solving questions

Challenge 1

PS **1** Look at the items and their weights.

A	B	C	D
20 g	100 g	Over 500 g	200 g

a) Which item weighs the least? Circle the correct letter.

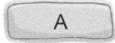

1 mark

b) Which item weighs the most? Circle the correct letter.

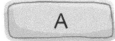

1 mark

c) Write the items in **letter** order from **lightest** to **heaviest**.

_____ _____ _____ _____

2 marks

Marks.......... /4

Challenge 2

PS **1** These bottles contain different amounts of juice.

A   B   C

a) Which bottle has the highest volume of juice?
Circle the correct letter.   A   B   C

b) Which bottle has the least amount?
Circle the correct letter.   A   B   C

Measuring Weight and Capacity

c) Put the bottles in order from **smallest** to **largest** volume.

_____ _____ _____

d) Bottle A contains 2 litres of juice. Which bottle contains about 1 litre of juice? Circle the correct letter.   **A   B   C**

e) If all three bottles were full, how much would they contain in total? Tick (✓) your answer.

3 litres ☐   6 litres ☐   10 litres ☐

5 marks

Marks.......... /5

Challenge 3

PS **1** Ali and her friends have been collecting cherries.

Fred = 1 kg   Ali = 2 kg   Sue = 2 kg   Ruby = 3 kg

a) Who collected the **most**? _____

1 mark

b) Who collected the **least**? _____

1 mark

c) Which two of the friends collected **equal** amounts?

_____ and _____

1 mark

d) Who collected the **same amount** as Ali and Fred added together? _____

1 mark

e) Put the weights in order from **heaviest** to **lightest**.

_____ _____ _____ _____

2 marks

Marks.......... /6

Total marks .... /15   How am I doing? 😊 😐 😣

Comparing Measurements

Challenge 1

PS | **1** | Lukas, Bella and Enrica planted three sunflowers. They have all started to grow.

A = 20 cm   B = 15 cm   C = 30 cm

a) Lukas's sunflower measures 20 cm. Is Lukas's flower

A, B or C? _____

b) Bella's sunflower measures 30 cm. Is Bella's flower

A, B or C? _____

c) Enrica's sunflower measures 15 cm. Which sunflower is

Enrica's? _____

d) Put the flowers in order from **shortest** to **tallest**.

shortest | | | | tallest

4 marks

Marks.......... /4

Challenge 2

PS | **1** | Cath wants to fill a paddling pool for her pet dog Hogan.

a) Cath has a 1 litre jug and the pool holds 10 litres.

How many jugs of water will she need? _____

Comparing Measurements

b) Cath's friend Gill has a pool that holds twice the volume of water. How many litres of water will Gill's pool hold? _____ litres

c) If you combined both pools, what would be the volume of water? _____ litres

d) How many 10 litre buckets would that be? _____

4 marks

Marks.......... /4

Challenge 3

  1 Ade has entered a pumpkin competition.

A = 5 kg　　**B = 4 kg**　　**C = 1 kg**

a) Ade's pumpkin weighs 4 kg.

Is Ade's pumpkin A, B or C? _____

b) Katie's pumpkin weighs 1 kg more than Ade's.

Katie's pumpkin weighs _____ kg.

c) Pumpkin C belongs to Chang.

How much does Chang's pumpkin weigh? _____ kg

d) Which pumpkins are equal to Katie's when added

together? _____ and _____

4 marks

Marks.......... /4

Total marks .... /12　　　　How am I doing?

Measuring Time

PS Problem-solving questions

Challenge 1

 **1** Draw lines to match the correct times to the clocks.

a) b) c) d) e)

5 o'clock 2 o'clock half past 8 3 o'clock half past 4

5 marks

Marks.......... /5

Challenge 2

1 Draw the minute and hour hands on these blank clock faces to show the times.

a) 1 o'clock b) 4 o'clock c) Half past 2

d) Half past 5 e) 7 o'clock f) Half past 11

6 marks

Marks.......... /6

Measuring Time

Challenge 3

PS **1** Use the grid to help you answer these questions.

Monday	Tuesday	Wednesday	Thursday
Friday	Saturday	Sunday	

a) Which day is 1 day after Tuesday? _____

b) Which day is 3 days before Friday? _____

c) If tomorrow is Thursday and yesterday was Tuesday,

what day is it today? _____

3 marks

2 Look at the months – they are not in the right order. Write them in the correct order.

January	March	June	February
April	December	August	October
July	November	May	September

_____

_____

_____

_____

_____

_____

3 marks

Marks.......... /6

Total marks .... /17

How am I doing?

Time Problems

Challenge 1

  1 Each day Lucy takes her dog Peg out for a walk. She sets off at 11 o'clock (shown on clock **A**) and walks for 2 hours.

A　　　　　**B**

a) Draw the time that Lucy and Peg get back home from their walk on clock B.

b) How many hours would Lucy and Peg walk in two days? _____ hours

c) What time is it when Lucy and Peg have been walking for 1 hour? _____ o'clock

3 marks

Marks........../3

Challenge 2

1 Put these events in order from the earliest to the latest.

  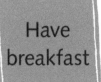

Have lunch	Go to bed	Wake up	Go to school	Have breakfast
A	**B**	**C**	**D**	**E**

Write the letters to show the correct order.

_____ _____ _____ _____ _____

3 marks

Time Problems

PS **2** Choose words from the cards to make these sentences true.

| first | after | next | before |

a) Squeeze on the toothpaste _____ you brush your teeth.

b) Wash the pots _____ supper.

c) _____ get dressed and then put on your shoes.

d) First pour in the juice, _____ the water.

4 marks

Marks.......... /7

Challenge 3

PS **1** Choose words from the shapes to make the sentences correct. Be careful, not all of the words are needed!

| morning | evening | today | tomorrow |

a) If it is the twelfth now, then _____ is the thirteenth.

b) I always wake up early in the _____.

c) Every _____ after school I play with my friends.

3 marks

PS **2** Circle the correct words.

a) 10 years is a **century** / **decade**.

b) There are 31 **days** / **weeks** in most months.

2 marks

Marks.......... /5

Total marks .... /15   How am I doing? ☺ 😐 ☹

Standard Units of Money

1 Look at the coins below.

a) How many coins are there in total? _____

b) What is the total value of the coins? _____p

2 marks

2 Look at the coins below.

a) How many coins are there in total? _____

b) What is the total value of the coins? _____p

2 marks

3 Look at the coins below.

a) How many coins are there in total? _____

b) What is the total value of the coins? _____p

2 marks

4 Put the coins in order from **least** valuable to **most** valuable.

least valuable   **most valuable**

6 marks

Marks.........../12

80

Standard Units of Money

1 Write this money in order from **least** valuable to **most** valuable.

   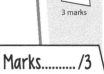

least
valuable

most
valuable

3 marks

Marks.........../3

1 Paulo has sorted his money into groups of the same coins. What is the value of each group of coins?

a)
 _____ p

b)    _____ p

c)   _____ p

d)
 _____ p

4 marks

Marks........./4

Total marks ..../19   How am I doing?

Money Problems

Challenge 1

PS **1** Josh has been to a garage sale and now he's ready for a day on the beach!

a) Josh bought 3 flags. They were 2p each.

How much did he spend? _____

b) The bucket and spade were 10p each.
How much did Josh spend on them in total? _____

c) Josh also bought a ball and sunglasses for 5p each.
How much did he spend on them in total?

_____

3 marks

Marks........../3

Challenge 2

PS **1** Solve these money problems.

Apple = 10p   **Banana**   **Lemon = 10p**   **Lime**

a) One apple costs 10p. How much would 2 apples be?

_____p

b) Adam buys 2 bananas for 40p. How much does each banana cost? _____p

Money Problems

c) Lemons are 10p each. How much would 3 lemons be?

_____p

d) Limes are £1 per bag. Eli has £2. How many bags can

she buy? _____

4 marks

Marks.......... /4

Challenge 3

PS **1** Mary has sorted her money into coins of the same type. Find the total of each set of Mary's coins.

a) Mary has **four** 10p coins.

The total amount is _____p.

b) She has **eight** 2p coins. They total _____p.

c) Mary has **ten** 5p coins. The total amount she has in

5p coins is _____p.

d) She has **two** 20p coins. Mary has _____p in 20p coins.

e) Mary has **two** 50p coins. Circle the amount she has in 50p coins.

£1   £2   £4

5 marks

Marks.......... /5

Total marks .... /12   How am I doing?

2-D Shapes

Challenge 1

1 2-D shapes are also known as **flat shapes**.
Choosing from the given words, write the correct name next to each shape.

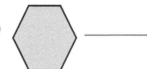

| hexagon | triangle | pentagon | circle | rectangle |

a) _____

b) _____

c) _____

d) _____

e) _____

5 marks

2 Find the number of each shape below.

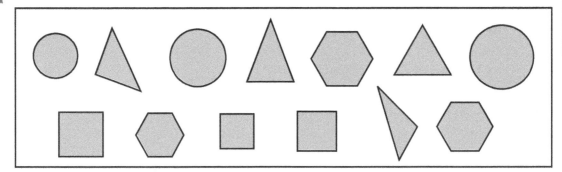

a) There are _____ triangles.

b) There are _____ squares.

c) There are _____ hexagons.

d) There are _____ circles.

e) How many shapes are there in total? _____

5 marks

Marks.........../10

84

2-D Shapes

Challenge 2

1 A triangle has three straight sides, but there are many different types of triangle. Draw five triangles and make each one different.

5 marks

Marks.......... /5

Challenge 3

1 Name three items that are usually rectangular. For example, a smartphone.

a) _____

b) _____

c) _____

3 marks

2 Circle the shapes that are **not** rectangles.

| A | B | C | D | E |

2 marks

Marks.......... /5

Total marks .... /20   How am I doing?

3-D Shapes

1 3-D shapes are also known as **solid shapes**.
Draw lines to match each 3-D shape to its name.

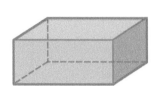
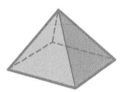
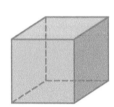

| cube | pyramid | cylinder | cuboid |

4 marks

2 Count the number of each 3-D shape.

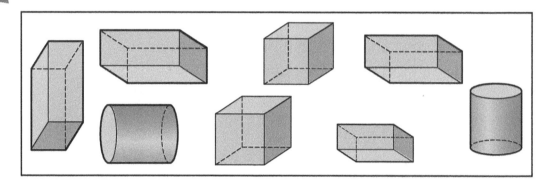

a) There are _____ cubes.

b) There are _____ cylinders.

c) There are _____ cuboids.

3 marks

Marks.........../7

3-D Shapes

Challenge 2

1 Draw five different cylinders.

<table>
<tr><td></td><td></td><td></td><td></td><td></td></tr>
</table>

5 marks

Marks.......... /5

Challenge 3

1 Name four items that are usually cuboid in shape. For example, a washing machine.

a) _____

b) _____

c) _____

d) _____

4 marks

Marks.......... /4

Total marks .... /16   How am I doing?

Different Shapes

Challenge 1

1 Draw five different 2-D shapes in the boxes below.

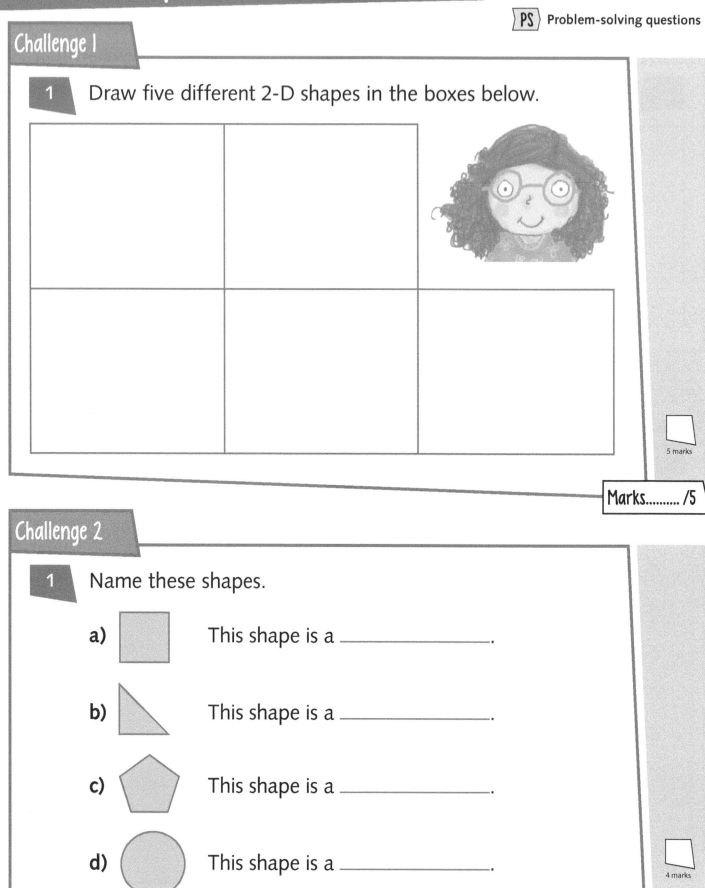

5 marks

Marks.........../5

Challenge 2

1 Name these shapes.

a) This shape is a _____.

b) This shape is a _____.

c) This shape is a _____.

d) This shape is a _____.

4 marks

Marks.........../4

Different Shapes

Challenge 3

PS | 1 | Everyday objects can resemble 3-D (solid) shapes. Draw lines to match each item to the name of the 3-D shape that it looks like.

a)

cube

b)

sphere

c)

cylinder

d)

cuboid

4 marks

Marks.......... /4

Total marks .... /13   How am I doing?

Progress Test 3

1. Look at the fish. There are 12 in total.

   a) Draw a blue ring around half of the fish.

   b) Draw a red ring around $\frac{1}{4}$ of the fish.

2 marks

2. Answer these divisions. You could use objects to help you.

   a) 6 shared by 2 = _____   **b)** 12 shared by 2 = _____

   c) 10 shared by 2 = _____   **d)** 8 shared by 2 = _____

   e) 14 shared by 2 = _____   **f)** 20 shared by 2 = _____

6 marks

PS 3. This playful kitten has 4 legs.

   a) How many legs would 3 playful kittens have? _____

   b) How many legs would 5 kittens have?   _____

2 marks

PS 4. Bess is collecting cherries.

   a) Each bowl holds 20 cherries. Bess collects
          2 bowlfuls.

   How many cherries has she collected? _____

   b) Bess puts her cherries into groups of 10.

   How many groups of 10 does she have? _____

   c) Bess shares all of her cherries with her brother Bob.

   How many cherries do they get each? _____

3 marks

PS 5. Answer these capacity questions.

a) If the capacity of one jug is 3 litres, what is the capacity of

2 jugs? _____ litres

b) Mila buys 5 litres of water.

How many 1 litre bottles can she fill? _____

c) David needs to fill his 20 litre fish tank.
He has a 2 litre container.

How many containers will he use to fill the tank? _____

d) Jason has 100 ml of ice cream.

How many 10 ml scoops can he serve? _____

4 marks

6. Find $\frac{1}{2}$ of these groups of penguins.

a) 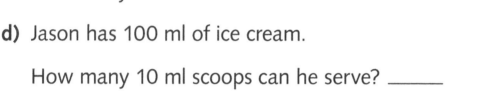 $\frac{1}{2}$ = _____

b)  $\frac{1}{2}$ = _____

2 marks

7. Choose the correct words from the boxes to describe each pair
of numbers.

| is less than | is greater than | is equal to |

a) 10 _____ 12   **b)** 20 _____ 18

c) 20 _____ 20   **d)** 8 _____ 16

4 marks

PS **8.** Answer these questions about length.

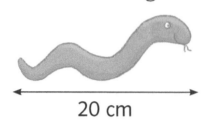

20 cm

a) The snake measures exactly 20 cm. What does half of the snake measure? _____ cm

b) What would be the total length of 2 snakes? _____ cm

c) How many snakes would be needed to measure 1 m (or 100 cm)? _____

3 marks

9. Add the times to the clocks.

a) 2 o'clock   b) Half past 10   c) 8 o'clock

d) Half past 7   e) 12 o'clock

5 marks

10. Put the weights in order from **lightest** to **heaviest**.

| 11 kg | 2 kg | 10 kg | 1 kg |

lightest | | | | | heaviest

2 marks

11. Look at these numbers.

   a) Use the numbers to make the highest value two-digit

   number you can. _____

   b) What is the lowest value two-digit number you can make

   with these numbers? _____

12. Divide the zebras into equal groups.

   a)    = _____ groups of 2

   b)

   = _____ groups of 2

13. Name these 3-D shapes.

   a)    This 3-D shape is a _____.

   b) 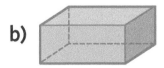   This 3-D shape is a _____.

Marks........ /39

Top, Middle and Bottom

Challenge 1

1 Here is a set of squares.

a) Colour the top square red.

b) Colour the bottom square blue.

c) Colour the middle square green.

3 marks

2 Look at the set of circles.
Answer the questions using these words:

stripes   squares   spots

a) What pattern is in the middle circle?

_____

b) How is the top circle filled? _____

c) What pattern does the bottom circle contain?

_____

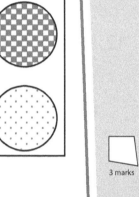

3 marks

Marks.......... /6

Challenge 2

1 Look at the shapes.

a) Which shape is at the bottom?

_____

b) Which shape is in the middle?

_____

c) What is the name of the top shape?

_____

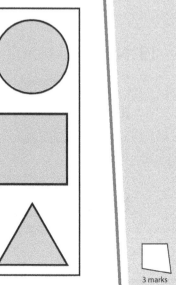

3 marks

Marks.......... /3

Top, Middle and Bottom

1 Look at the large squares on the right.

a) Draw a funny face in the middle square.

b) Draw a star in the top square.

c) Draw an apple in the bottom square.

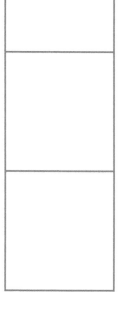

3 marks

2 Now look at the small squares on the right.

a) Colour the top 2 squares yellow.

b) Colour the middle 2 squares red.

c) Colour the bottom 2 squares green.

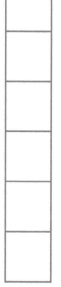

3 marks

Marks.......... /6

Total marks .... /15

How am I doing?

95

Around, Inside and Outside

Challenge 1

1 **a)** Draw dots around the outside of this square.

b) Draw four circles inside the square.

2 marks

2 **a)** Draw four Xs around the circle.

b) Put dots inside the circle.

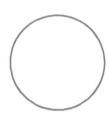

2 marks

Marks.......... /4

Challenge 2

1 Look at the shapes on the right.

a) Which two shapes are on the inside?

_____ and _____.

b) Which shape is outside both of the other shapes?

_____

2 marks

2 Look at the shapes on the right.

a) Which two shapes is the square inside?

_____ and _____.

b) Which shape is on the outside?

_____

2 marks

Marks.......... /4

Around, Inside and Outside

 Look at the square below.

a) Draw four circles around the square.

b) Draw a dot inside each circle.

c) Draw a triangle inside the square.

d) Draw an X inside the triangle.

4 marks

Marks.......... /4

Total marks .... /12   How am I doing?

Describing Positions

Challenge 1

PS **1** Charlie the cat is a little lost! He needs help to make the right choice.

Dog   Charlie   Tree

a) If Charlie turns to your right, what will he see?

_____

b) If he turns to your left, what will he see? _____

c) Which way do you think Charlie should go? Why?

_____

_____

3 marks

Marks.........../3

Challenge 2

PS **1** Help Agneta with her directions! Look at her position in the grid and then answer the questions on the next page.

	Pizza	
Ice cream	Agneta	Banana
	Lolly	

Describing Positions

a) What is in front of Agneta? _____

b) What is behind Agneta? _____

c) What is to the right of Agneta? _____

d) What is to the left of Agneta? _____

4 marks

Marks.......... /4

PS | 1 | Draw the items below in the correct places on the grid.

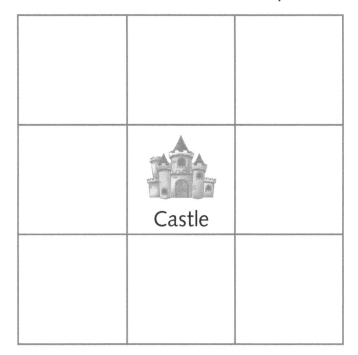

Castle

a) Draw a star **above** the castle.

b) Draw a key **below** the castle.

c) Draw a coin on the **right** of the castle.

d) Draw a crown to the **left** of the castle.

4 marks

Marks.......... /4

Total marks .... /11   How am I doing?

Left and Right Turns

PS Problem-solving questions

Challenge 1

1 Look at the image below.

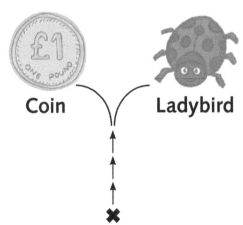

Coin   Ladybird

a) Start at the cross and follow the arrows.
Then make a right turn.
What is there?   _____

b) What would be there if you made
a left turn instead?   _____

2 marks

PS 2 Is this sentence **true** or **false**? Circle your answer.

Clockwise is a turn to the right.   **True**   **False**

1 mark

Marks............/3

Challenge 2

1 Look at the shaded arrow on the right – it is pointing
straight up.
Choose from the white arrows **A**, **B** and **C** below to
answer the questions on the next page.

A   B   C

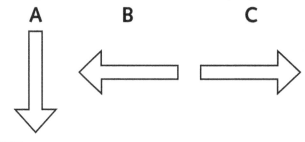

Left and Right Turns

a) If the shaded arrow turned a $\frac{1}{4}$ clockwise turn, which white arrow would it be? _____

b) If the shaded arrow turned a half turn to the left or right, which white arrow would show its final position? _____

c) If the shaded arrow turned a quarter turn to the left, which white arrow would it be? _____

3 marks

Marks............/3

Challenge 3

PS 1 Write down the directions to the coin. Follow the path as shown by the arrows.

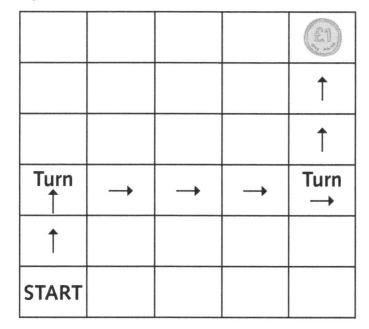

a) Move _____ blocks. **b)** Turn _____ .

c) Move _____ blocks. **d)** Turn _____ .

e) Move _____ blocks.

5 marks

Marks............/5

Total marks .... /11   How am I doing?

More Position and Direction

Challenge 1

PS〉 **1** Help Charlie to make his journey.
Choose the correct word to pass each obstacle.
The arrows help you choose the correct words.

Bridge

Stepping stones

Tree

Arch

Pipe

| under | over | across | around | through |

a) Charlie should go _____ the bridge.

b) He should go _____ the stepping stones.

c) Charlie needs to go _____ the tree.

d) He should go _____ the arch.

e) Charlie needs to go _____ the pipe.

5 marks

Marks.......... /5

More Position and Direction

Challenge 2

 1 Use words from the boxes to describe Anna's challenge.

| up | down | under | across |

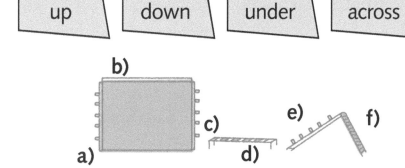

Anna went **a)** _____ the climbing wall,

b) _____ the balance, **c)** _____ the

climbing wall, **d)** _____ the net, **e)** _____

the ramp and **f)** _____ the ladder.

6 marks

Marks.......... /6

Challenge 3

PS **1** Sam wanted a fun photo. Help her to describe it to her friends.

a) What is on top of Sam's head? _____

1 mark

b) What is Sam standing on top of?

_____

1 mark

c) Sam is between the _____

and the _____.

2 marks

Marks.......... /4

Total marks .... /15   How am I doing?

1. Write each set of numbers in order from **least** to **greatest**.

a)  3 | 13 | 11 | 2

b)  17 | 7 | 15 | 27

c) 1 | 9 | 6 | 4

6 marks

2. Write the addition and subtraction families for these sets of numbers.

a) 3 2 5

_____ _____ _____ _____

b) 6 4 10

_____ _____ _____ _____

c) 5 6 11

_____ _____ _____ _____

3 marks

PS **3.** Flowers are sold in bunches of 5.
There are 4 bunches of flowers.
How many flowers are there altogether?

_____

1 mark

104

4. Use division to halve each number.

Example: Half of 10 is <u>5</u>

a) Half of 16 is _____

b) Half of 8 is _____

c) Half of 22 is _____

d) Half of 18 is _____

e) Half of 30 is _____

5 marks

5. Look at the **shaded** skittle on the left. Circle the skittle that would be a **quarter turn clockwise**.

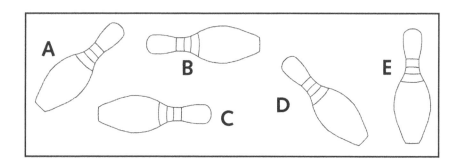

A

B

C

D

E

1 mark

6. Write the answers to these additions.

a) $2 + 2 + 2 =$ _____

b) $10 + 10 + 10 =$ _____

c) $5 + 5 + 5 + 5 =$ _____

3 marks

PS Problem-solving questions

PS **7.** Solve the following number problems.

a) Jenny has 6 rings. She loses 2 rings. How many rings does she now have? ______________________

b) Hameed sets out to school at 8 o'clock in the morning. He arrives at 9 o'clock morning time. How many hours did the journey take? Circle the correct answer.

1 hour   2 hours   3 hours

2 marks

8. Draw lines to match the words to the numerals.

sixteen	9
seven	8
nine	16
eight	7

4 marks

9. a) Draw a line 6 cm long.

b) Draw a line 3 cm long.

2 marks

PS **10.** Look at the grid below.

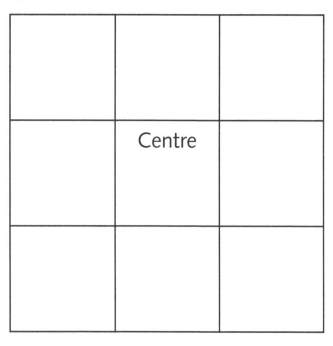

	Centre	

Add the following to the grid.

a) Draw a triangle in the centre square.

b) Draw a tick (✓) **above** the triangle.

c) Draw a square **below** the triangle.

d) Draw a circle in one of the squares to the **right** of the triangle.

e) Draw a cross (✗) in one of the squares to the **left** of the triangle.

 5 marks

11. Shade or colour in the fractions of each circle.

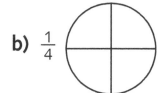 a) $\frac{1}{2}$ 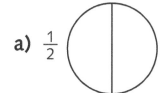   b) $\frac{1}{4}$

 2 marks

Marks........ /34

107

Notes

Notes

Notes

Answers

Pages 4–11
Starter Test
1 a) 1, **2**, 3, 4, **5**, 6, **7**, 8, **9**, 10
   b) 2, **4**, 6, **8**, 10, **12**, 14, **16**, **18**, 20
   c) 0, 5, 10, 15, **20**, 25, 30
2 a) **3**, 4, **5**
   b) **8**, 9, **10**
   c) 5, 6, 7
   d) 7, 8, **9**
   e) 4, 5, **6**
   f) 6, 7, **8**
3 Shapes **A** and **C** ticked
4 a) 5   b) 4
   c) 6   d) 10
5 a) True
   b) False
   c) True
6 a) 7   b) 15
   c) 4   d) 20
   e) 17   f) 0
7 a) 6 eggs
   b) 2 + 2 + 2 = 6
8 a) Twelve   b) Ten
   c) Seven   d) Six
   e) Zero   f) Eighteen
   g) Eleven   h) Five
   i) Nine   j) Sixteen
9 a) 10 apples
   b) 5 + 5 = 10
   c) 15 apples
   d) 5 + 5 + 5 = 15
10 a) 3 + 1 = 5 circled
   b) 2 + 2 = 6 circled
   c) 14 – 5 = 10 circled
11 a)

   b)

   c)

12 a) Snake C   b) Shorter
   c) Snake D
   d) C, A, B, D **(1 mark for 2 snakes in correct order; 2 marks for all correct)**

13 a) 10
   b) 20 – 10 = 10 or half of 20 = 10
14 a) 16   b) 20
   c) 30   d) 24
   e) 25   f) 60
   g) 50   h) 18
   i) 40   j) 30
15 a) 6   b) 4
   c) 9
16 a) Sequences **B** and **C** ticked
   b) Sequence **A** is counting back/down in 1s
      Sequence **B** is counting forwards/up in 2s
   c) Sequence **C** ticked
   d) Because it is the only sequence counting in 2s. The others are counting in steps of 5.
17 a) 1 ten, 5 ones
   b) 1 ten, 6 ones
   c) 2 tens, 9 ones
18 a) 6
   b) 8
   c) 10
   d) 12
19 a) 7
   b) 8
20 a) 17
   b) 15
   c) 11
21 a) Pencil ⟶ less than 20 g
      Apple ⟶ about 100 g
      Chair ⟶ more than 10 kg
      Potatoes ⟶ about 3 kg
   b) The pencil

Pages 12–13
Challenge 1
1 a) 11 ticked
   b) 2, 3, 6, 11 **(1 mark for 2 numbers in correct order; 2 marks for all correct)**
2 a) 32 ticked
   b) 14, 18, 22, 24, 32 **(1 mark for 2 numbers in correct order; 2 marks for 3 correct; 3 marks for all correct)**
Challenge 2
1 a) 10 **or** 1 ten   b) 3 **or** 3 ones
   c) 20 **or** 2 tens   d) 40 **or** 4 tens
2 a) 2 tens, 6 ones
   b) 26 = 20 + 6
3 **3**, 4, **5**, **6**, 7, 8, **9**, 10, **11**, **12** 13, 14, **15**, **16**, **17**, 18, **19**, **20**, 21, 22
   (1 mark for 6–10 numbers correct; 2 marks for all correct)

113

Answers

Challenge 3

1

1	2	3	4	5	6	7	8	9	10
11	12	13	14	15	16	17	18	19	20
21	22	23	24	25	26	27	28	29	30
31	32	33	34	35	36	37	38	39	40
41	42	43	44	45	46	47	48	49	50
51	52	53	54	55	56	57	58	59	60
61	62	63	64	65	66	67	68	69	70
71	72	73	74	75	76	77	78	79	80
81	82	83	84	85	86	87	88	89	90
91	92	93	94	95	96	97	98	99	100

(1 mark for each correct row)

Pages 14–15
Challenge 1
1 a) 14　　　　b) 9
　　c) 7　　　　d) 5
2 a) 7　　　　b) 10
　　c) 12　　　　d) 15

Challenge 2
1 7 birds
2 17 stickers
3 **1**, 2, **3**, 4, **5**, 6, **7**, 8, **9**, 10

Challenge 3
1 4, 5, 13, 14, 15, 19, 20 **(1 mark for 2 numbers in correct order; 2 marks for 3 correct; 3 marks for 4 correct; 4 marks for 5 correct; 5 marks for all correct)**
2 a)

1	2	3
11	12	13
21	22	23

　　b)

23	24	25
33	34	35
43	44	45

Pages 16–17
Challenge 1
1 a) 2, 4, 6, 8, **10**, **12**, **14**, **16**
　　b) 15, 13, 11, **9**, **7**, **5**, **3**
2 a) Two
　　b) Five

Challenge 2
1 a) 15, **20**, 25, 30, **35**, 40
　　b) 20, **25**, 30, 35, **40**, 45
2 a) Backwards
　　b) Forwards

Challenge 3
1 a) **5**, 10, **15**, 20
　　b) 2, **4**, **6**, 8
2 True

Pages 18–19
Challenge 1
1 30
2 2
3

10	20	30	40
50	60	70	80

(1 mark for 3–5 numbers correct; 2 marks for all correct)

4 40

Challenge 2
1 10, 20, 30, 40, 50 **(1 mark for 2 numbers in correct order; 2 marks for 3 correct; 3 marks for all correct)**
2 40, 50, 60, 70, 80 **(1 mark for 2 numbers in correct order; 2 marks for 3 correct; 3 marks for all correct)**

Challenge 3
1 a) 3　　　　b) 6
　　c) 5　　　　d) 3

Pages 20–21
Challenge 1
1 a) **14**, 15, **16**
　　b) **17**, 18, **19**
　　c) **8**, 9, **10**
　　d) **18**, 19, **20**
　　e) **16**, 17, **18**
2 a) 7　　　　b) 9
　　c) 16

Challenge 2
1 a) 7　　　　b) 17
　　c) 25　　　　d) 15
2 a) 16　　　　b) 29
　　c) 11　　　　d) 24
3 19

Challenge 3
1 a) 4　　　　b) 16
　　c) 30　　　　d) 22
2 a) 10　　　　b) 10
3 a) 30
　　b) 50, 20 + 30 = 50 **or** 20 + 20 + 10 = 50

Pages 22–23
Challenge 1
1 a) 21   b) 15
   c) 29   d) 43
2 a)
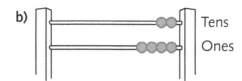
   b)
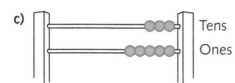
   c)
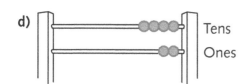
   d)

Challenge 2
1 a) 10 + 2   b) 20 + 5
   c) 30 + 8   d) 40 + 6
   e) 50 + 3
2 a) 25   b) 31
   c) 42   d) 37

Challenge 3
1 Accept 12, 13 or 14
2 Accept 16, 17, 18 or 19
3 a) 1 ten, 4 ones
   b) 2 tens, 4 ones
   c) 3 tens, 6 ones
   d) 4 tens, 8 ones
4 2, 15, 16, 28 **(1 mark for 2 numbers in correct order; 2 marks for all correct)**

Pages 24–25
Challenge 1
1 a) Any pair where the first number is less than the second, e.g. 17 and 20
   b) Any pair where the first number is greater than the second, e.g. 20 and 17
   c) 1 and 1 **or** 20 and 20
2 a) Any pair where the first number is greater than the second
   b) Any pair where the first number is less than the second
   c) Any pair of equal numbers

Challenge 2
1 a) is less than   b) is equal to
   c) is greater than
2 a) is less than   b) is greater than
   c) is equal to

Challenge 3
1 a) equal   b) fewer
   c) more   d) fewer

Pages 26–27
Challenge 1
1 a) Double 5 is 10, half of 10 = 5
   b) Double 7 is 14, half of 14 = 7
2 a) 2 + 2 = 4   b) 6 + 6 = 12
   c) 10 + 10 = 20   d) 7 + 7 = 14

Challenge 2
1 a) 5 + 10 = 15, 10 + 5 = 15, 15 − 5 = 10, 15 − 10 = 5
   b) 3 + 7 = 10, 7 + 3 = 10, 10 − 3 = 7, 10 − 7 = 3
   c) 2 + 3 = 5, 3 + 2 = 5, 5 − 2 = 3, 5 − 3 = 2
   d) 20 + 10 = 30, 10 + 20 = 30, 30 − 20 = 10, 30 − 10 = 20

Challenge 3
1 a) 6   b) 6
   c) 3   d) 6
2 a) 12   b) 13
   c) 5   d) 11

Pages 28–29
Challenge 1
1 8 − 6 = 2
2 10 + 7 = 17
3 12 − 3 = 9

Challenge 2
1 a) 30
   b) 10 + 10 + 10 = 30
2 a) 4
   b) 12 − 8 = 4
3 16

Challenge 3
1 a) 5 + 3 + 1 = 9 **or** any other sequence of the given numbers = 9
   b) 3 + 4 + 2 = 9 **or** any other sequence of the given numbers = 9
   c) 6 + 1 + 5 = 12 **or** any other sequence of the given numbers = 12

Pages 30–31
Challenge 1
1 1 + 9, 9 + 1, 8 + 2, 2 + 8, 3 + 7, 7 + 3, 4 + 6, 6 + 4, 5 + 5, 0 + 10, 10 + 0

Answers

Challenge 2
1. a)–j) Any two-digit numbers using two different one-digit numbers available.
2. a) 13
   b) 14
   c) 16
   d) 11
   e) 18

Challenge 3
1. Any five subtractions that correctly result in 10 as the answer.
2. 14, **15**, 16, **17**, **18**, 19, **20**

Pages 32–33
Challenge 1
1. a) 6   b) 4
   c) 2   d) 1
   e) 4   f) 3
   g) 12   h) 10
   i) 10

Challenge 2
1.

2	3	4	5	6
7	8	9	10	11
12	13	14	15	16
17	18	19	20	21
22	23	24	25	26

(1 mark for each correct row)

2. a) +   b) +
   c) −   d) +
   e) −

Challenge 3
1. a) 8p   b) 15p
   c) 17p (accept 2p more than answer to part **b**))
   d) 23p (accept sum of answers to parts **a**) and **b**))
   e) 4p

Pages 34–35
Challenge 1
1. a) False   b) True
   c) False
2. a) 19   b) 21
   c) 15

Challenge 2
1. a) 15   b) 15, 16, 17, 18, 19, 20
   c) 22   d) 5

Challenge 3
1. a) ✓   b) ✗
   c) ✓

2. a) 7   b) 12 − 5 = 7
3. a) True   b) False
4. a) 20
   b) Accept any four subtractions that leave 5

Pages 36–39
Progress Test 1
1. a) 10
   b) 8
   c) 20
   d) 18
2. a) 1   b) 2
   c) 5   d) 10
3. Any eight two-digit numbers using the digits 1, 2, 3, 4, 5 **(1 mark for 1–2 correct numbers, 2 marks for 3–4 numbers, 3 marks for 5–6 numbers, 4 marks for 6 numbers, 5 marks for all 8 numbers)**
4. a) 7
   b) 4 (accept 3 less than answer to part **a**))
   c) 10 (accept 6 more than answer to part **b**))
5. a) 3 + 9 = 12
   b) 8 + 2 = 10
   c) 4 + 7 = 11
   d) 10 + 6 = 16
6. 20 circled
7. a) 7   b) 14
   c) 16   d) 20
8. a) 21
   b) 15
   c) 11
9. a) 4
   b) 5
   c) 15
   d) 11
10. a) Five
    b) Seven
    c) Nine
    d) Two
    e) Twelve
11.

1	5	4
1	3	6
8	2	0

(1 mark for each correct row)

12. a) 12
    b) 98
    c) Any odd number possible from the given numbers.
    d) Any even number possible from the given numbers.

13 a) 17 + 3 = 20, 3 + 17 = 20, 20 − 17 = 3,
20 − 3 = 17

b) 14 + 5 = 19, 5 + 14 = 19, 19 − 14 = 5,
19 − 5 = 14

c) 11 + 4 = 15, 4 + 11 = 15, 15 − 11 = 4,
15 − 4 = 11

14 a) Stars put into groups of 5

b) 13

Pages 40–41
Challenge 1
1 a) 10

b) 5 sets of 2 = 10

2 a) 3 sets of 2 = 6

b) 2 sets of 5 = 10

c) 3 sets of 10 = 30

Challenge 2
1 5 sets of 2 = 10, 2 sets of 5 = 10

2 (1 mark for each correct number of rows,
1 mark for each correct number of columns)

a)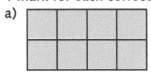

b)

c)

Challenge 3
1

Addition	Number of sets
2 + 2 + 2 + 2 + 2 = 10	5 sets of 2
5 + 5 + 5 = 15	3 sets of 5
10 + 10 + 10 = 30	**3 sets of 10**
5 + 5 + 5 + 5 = 20	4 sets of 5
10 + 10 + 10 + 10 + 10 = 50	5 sets of 10
2 + 2 + 2 + 2 + 2 + 2 = 12	**6 sets of 2**

Pages 42–43
Challenge 1
1 a) 2   b) 3

c) 5   d) 8

Challenge 2
1 4

2 6

3 3

Challenge 3
1

Number of items	Number of people	They each get
6	2	3
14	2	7
16	2	8
18	2	9
20	2	10

Pages 44–45
Challenge 1
1

2 times 1 = 2	5 times 1 = 5	10 times 1 = 10
2 times 2 = 4	**5 times 2 = 10**	**10 times 2 = 20**
2 times 3 = 6	5 times 3 = 15	10 times 3 = 30
2 times 4 = 8	**5 times 4 = 20**	**10 times 4 = 40**
2 times 5 = 10	5 times 5 = 25	10 times 5 = 50
2 times 6 = 12	5 times 6 = 30	**10 times 6 = 60**

Challenge 2
1

1	2	3	4	5
6	7	8	9	10
11	12	13	14	15
16	17	18	19	20

(1 mark for each correct row)

2 a) False

b) True

Challenge 3
1 a) 2 times 4 = 8

b) 10 times 1 = 10

2 a) 3 times 2 = 6, 2 times 3 = 6

b) 5 times 2 = 10, 2 times 5 = 10

Answers

Pages 46–47
Challenge 1
1   6 shared by 2 = 3
2   a) 10 shared by 2 = 5 **or** 10 shared by 5 = 2
    b) 20 shared by 10 = 2 **or** 20 shared by 2 = 10
    c) 16 shared by 2 = 8 **or** 16 shared by 8 = 2
3   a) 20 shared by 2 = 10 **or** 20 shared by 10 = 2
    b) 8 shared by 2 = 4 **or** 8 shared by 4 = 2
    c) 12 shared by 6 = 2 **or** 12 shared by 2 = 6
Challenge 2
1   a) 10   b) 20 shared by 2 = 10
    c) 5   d) 20 shared by 4 = 5
Challenge 3
1   a) 2   b) 4   c) 10

Pages 48–49
Challenge 1
1   a) 10   b) 20
    c) 6
2   a) 2   b) 4   c) 5
Challenge 2
1   a) 10
    b) 5
2   a) 12
    b) 24
Challenge 3
1   a) 4   b) 2
    c) 10   d) 1

Pages 50–51
Challenge 1
1   a) 5   b) 10
2   a) 8   b) 4
3   a) 6   b) 15
Challenge 2
1   a) 10
    b) 5
    c) 2
2   a) 10
    b) 3
Challenge 3
1   a) 4 spots
    b) 20 boys
    c) 30 stickers

Pages 52–53
Challenge 1
1   a) 3   b) 6
    c) 10   d) Even
    e) Odd
Challenge 2
1   a) 4   b) 5
    c) 7

Challenge 3
1   a) 15, 20 − 5 = 15
    b) 5
    c) 10

Pages 54–55
Challenge 1
1   a) 2
    b) 3
    c) 6
Challenge 2
1   6
2   a) 4   b) 5
    c) 6
3   a) False
    b) True
Challenge 3
1   a) 8
    b) 4 groups of 2 = 8
2   a) 10
    b) 2 groups of 5 = 10
3   a) ✓
    b) ✗
    c) ✗
    d) ✓

Pages 56–57
Challenge 1
1   a)–d) Any one half of each circle shaded
2   4
Challenge 2
1   a)–d) Any one half of each square shaded
2   8
Challenge 3
1   a) ✓   b) ✗
    c) ✗   d) ✗
    e) ✗   f) ✓

Pages 58–59
Challenge 1
1   a)–d) Any one quarter of each shape shaded
2   a)–d) Any one quarter of each shape shaded
3   Shapes **A** and **C** ticked
Challenge 2
1   a) 1 quarter
    b) 3 quarters
    c) 2 quarters
    d) 4 quarters
    e) 0 quarters
2   a) 3 quarters
    b) 1 quarter
    c) 2 quarters
    d) 0 quarters
    e) 4 quarters

Challenge 3
1 a) ✗
    b) ✗
    c) ✗
    d) ✓
    e) ✓

Pages 60–61
Challenge 1
1 a) 3
    b) 4
    c) 5
    d) 6
    e) 2
Challenge 2
1 a) 4
    b) 8
    c) 4
Challenge 3
1

Items	Amount in $\frac{1}{4}$	Amount in $\frac{1}{2}$
4 stickers	1	2
20 buns	5	10
8 straws	2	4

Pages 62–63
Challenge 1
1 a) 1
    b) 3
    c) 1
    d) 2
2 $\frac{1}{2}$ — greatest

    $\frac{1}{4}$ — smallest
Challenge 2
1 a) 6
    b) 3
    c) 8
    d) 5
    e) 5
2 a) False
    b) True
    c) True
    d) False
    e) True
Challenge 3
1 a) 8
    b) 20
    c) 20
    d) 16

Pages 64–65
Challenge 1
1 a) 2
    b) 2
    c) 2
Challenge 2
1 a) 8
    b) 4
    c) 4
    d) 2
Challenge 3
1 a) 6 cm
    b) 3 cm
    c) 2 cm
    d) 4 cm

Pages 66–69
Progress Test 2
1 a) 9
    b) 15
    c) 11
    d) 11
2 a) 2 tens, 7 ones
    b) 1 ten, 9 ones
    c) 2 tens, 1 one
    d) 2 tens, 5 ones
    e) 1 ten, 5 ones
3 a) 10
    b) 5
    c) 20
4 a) 8
    b) 10
    c) 4
    d) 5
    e) 7
5 a) 4
    b) 10
    c) 20
    d) 16
    e) 14
6 a)

    b)

7 a) 4
    b) 10
    c) 20

Answers

8 0, 5, **10**, **15**, **20**, 25
9 **a)** **12**, 13, **14**
    b) **10**, 11, **12**
    c) **18**, 19, **20**
10 **a)** is greater than
     b) is less than
     c) is greater than
     d) is equal to
     e) is less than
11 **a)** 3 + 5 = 8, 5 + 3 = 8
     b) 3 + 2 + 1 = 6 and 3, 2 and 1 added in any different order = 6
12 **a)** 1
     b) 2
     c) 6
     d) 2

Pages 70–71
Challenge 1
1 **a)** Snake D   **b)** Snake A
    c) Shorter
    d) A, C, B, D **(1 mark for 2 snakes in correct order; 2 marks for all correct)**
    e) 15 cm
2 **a)** 3 m
    b) 2 m
    c) 2 m
    d) 2 m
Challenge 2
1 Chair ⟶ about 1 m
    Pencil ⟶ less than 1 m
    Laptop ⟶ less than 1 m
    Tree ⟶ about 3 m
    Castle ⟶ over 5 m
2 **a)** 3 cm
    b) 9 cm
Challenge 3
1 **a)** 1 cm, 2 cm, 3 cm, **4 cm**, **5 cm**, 6 cm
    b) 5 cm, **10 cm**, **15 cm**, **20 cm**, 25 cm, **30 cm**
    c) 3 cm, **5 cm**, 7 cm, 9 cm, **11 cm**, **13 cm**

Pages 72–73
Challenge 1
1 **a)** A circled
    b) C circled
    c) A, B, D, C **(1 mark for 2 items in correct order; 2 marks for all correct)**
Challenge 2
1 **a)** A circled
    b) B circled
    c) B, C, A
    d) C circled
    e) 6 litres ticked

Challenge 3
1 **a)** Ruby
    b) Fred
    c) Ali and Sue
    d) Ruby
    e) 3 kg (Ruby), 2 kg (Sue) and 2 kg (Ali) in either order, 1 kg (Fred) **(1 mark for 2 weights in correct order; 2 marks for all correct)**

Pages 74–75
Challenge 1
1 **a)** A
    b) C
    c) B
    d) B, A, C
Challenge 2
1 **a)** 10
    b) 20 litres
    c) 30 litres
    d) 3
Challenge 3
1 **a)** B   **b)** 5 kg
    c) 1 kg   **d)** B and C

Pages 76–77
Challenge 1
1 **a)** 2 o'clock
    b) half past 4
    c) 3 o'clock
    d) half past 8
    e) 5 o'clock
Challenge 2
1 **a)**

    b)

    c)

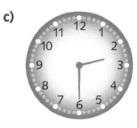

Answers

d)

e)

f)

Challenge 3
1 a) Wednesday
   b) Tuesday
   c) Wednesday
2 January, February, March, April, May, June, July, August, September, October, November, December **(1 mark for 4–7 months in correct order; 2 marks for 8–11 correct; 3 marks for all correct)**

Pages 78–79
Challenge 1
1 a) 1 o'clock drawn on clock B
   b) 4 hours
   c) 12 o'clock
Challenge 2
1 C, E, D, A, B **(1 mark for 2 events in correct order; 2 marks for 3 correct; 3 marks for all correct)**
2 a) before
   b) after
   c) first
   d) next
Challenge 3
1 a) tomorrow
   b) morning
   c) evening
2 a) decade circled
   b) days circled

Pages 80–81
Challenge 1
1 a) 7
   b) 7p
2 a) 5
   b) 10p
3 a) 4
   b) 20p
4 1p, 2p, 5p, 10p, 20p, 50p, £1, £2 **(1 mark for 2 coins in correct order; 2 marks for 3 correct; 3 marks for 4 correct; 4 marks for 5 correct; 5 marks for 6 correct; 6 marks for all correct)**
Challenge 2
1 £1, £2, £5, £10, £20 **(1 mark for 2 amounts in correct order; 2 marks for 3 correct; 3 marks for all correct)**
Challenge 3
1 a) 16p
   b) 35p
   c) 20p
   d) 10p

Pages 82–83
Challenge 1
1 a) 6p
   b) 20p
   c) 10p
Challenge 2
1 a) 20p
   b) 20p
   c) 30p
   d) 2
Challenge 3
1 a) 40p
   b) 16p
   c) 50p
   d) 40p
   e) £1 circled

Pages 84–85
Challenge 1
1 a) pentagon
   b) rectangle
   c) circle
   d) hexagon
   e) triangle
2 a) 4
   b) 3
   c) 3
   d) 3
   e) 13
Challenge 2
1 Accept drawings of any five different triangles

Answers

Challenge 3
1 a)–c) Accept any three rectangular items
2 Shapes **B** and **D** circled

Pages 86–87
Challenge 1

1 cylinder →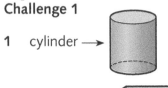

   cuboid →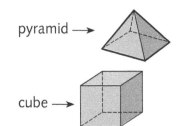

   pyramid →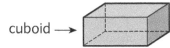

   cube →

2 a) 2
   b) 2
   c) 4

Challenge 2
1 Accept drawings of any five different cylinders

Challenge 3
1 a)–d) Accept any four items that are cuboids

Pages 88–89
Challenge 1
1 Accept drawings of any five different 2-D shapes

Challenge 2
1 a) square
   b) triangle
   c) pentagon
   d) circle

Challenge 3
1 a) cylinder
   b) sphere
   c) cuboid
   d) cube

Pages 90–93
Progress Test 3
1 a) 6 fish ringed in blue
   b) 3 fish ringed in red
2 a) 3
   b) 6
   c) 5
   d) 4
   e) 7
   f) 10

3 a) 12
   b) 20
4 a) 40
   b) 4
   c) 20
5 a) 6 litres
   b) 5
   c) 10
   d) 10
6 a) 3
   b) 4
7 a) is less than
   b) is greater than
   c) is equal to
   d) is less than
8 a) 10 cm
   b) 40 cm
   c) 5
9 a)

   b)

   c)

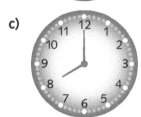

   d)

   e)

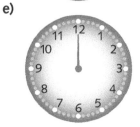

10 1 kg, 2 kg, 10 kg, 11 kg **(1 mark for 2 weights in correct order; 2 marks for all correct)**
11 **a)** 97
    b) 36
12 **a)** 2
    b) 3
13 **a)** cylinder
    b) cuboid

Pages 94–95
Challenge 1
1   **a)** Top square coloured red
    b) Bottom square coloured blue
    c) Middle square coloured green
2   **a)** Squares
    b) Stripes
    c) Spots
Challenge 2
1   **a)** Triangle
    b) Square
    c) Circle
Challenge 3
1   **a)** A funny face drawn in the middle square
    b) A star drawn in the top square
    c) An apple drawn in the bottom square
2   **a)** Top two squares coloured yellow
    b) Middle two squares coloured red
    c) Bottom two squares coloured green

Pages 96–97
Challenge 1
1   **a)** Dots drawn around the square
    b) Four circles drawn inside the square
2   **a)** Four Xs drawn around the circle
    b) Dots drawn inside the circle
Challenge 2
1   **a)** Triangle and circle
    b) Square
2   **a)** Circle and triangle
    b) Triangle
Challenge 3
1   **a)** Four circles drawn around the square
    b) A dot drawn inside each of the four circles
    c) A triangle drawn inside the square
    d) An X drawn inside the triangle

Pages 98–99
Challenge 1
1   **a)** The tree
    b) The dog
    c) Accept either left or right with a valid reason

Challenge 2
1   **a)** Pizza
    b) Lolly
    c) Banana
    d) Ice cream

Challenge 3
1

d) Crown	**a)** Star	**c)** Coin
d) Crown	Castle	**c)** Coin
d) Crown	**b)** Key	**c)** Coin

Accept any of the positions shown for the coin and crown

Pages 100–101
Challenge 1
1   **a)** The ladybird   **b)** The coin
2   True circled
Challenge 2
1   **a)** C   **b)** A
    c) B
Challenge 3
1   **a)** Forward (up) 2
    b) $\frac{1}{4}$ turn right **or** $\frac{1}{4}$ turn clockwise
    c) Forward (right) 4
    d) $\frac{1}{4}$ turn left
    e) Forward (up) 3

Pages 102–103
Challenge 1
1   **a)** over **or** across
    b) across **or** over
    c) around
    d) under
    e) through

Answers

Challenge 2
1 a) up   b) across
   c) down   d) under
   e) up **or** across   f) down

Challenge 3
1 a) A bird   b) A box
   c) The bird and the box

Pages 104–107
Progress Test 4
1 a) 2, 3, 11, 13 **(1 mark for 2 numbers in correct order; 2 marks for all correct)**
   b) 7, 15, 17, 27 **(1 mark for 2 numbers in correct order; 2 marks for all correct)**
   c) 1, 4, 6, 9 **(1 mark for 2 numbers in correct order; 2 marks for all correct)**
2 a) $3 + 2 = 5, 2 + 3 = 5, 5 - 3 = 2, 5 - 2 = 3$
   b) $6 + 4 = 10, 4 + 6 = 10, 10 - 6 = 4,$
      $10 - 4 = 6$
   c) $5 + 6 = 11, 6 + 5 = 11, 11 - 5 = 6,$
      $11 - 6 = 5$
3 20
4 a) 8   b) 4
   c) 11   d) 9
   e) 15
5 Skittle **C** circled
6 a) 6   b) 30
   c) 20
7 a) 4
   b) 1 hour circled

8 sixteen ⟶ 16
   seven ⟶ 7
   nine ⟶ 9
   eight ⟶ 8
9 a) Accept any line measuring between 5 cm and 7 cm
   b) Accept any line measuring between 2 cm and 4 cm

10

✗ e) Cross	✓ b) Tick	◯ d) Circle
✗ e) Cross	Centre △ a) Triangle	◯ d) Circle
✗ e) Cross	▢ c) Square	◯ d) Circle

Accept any of the positions shown for the circle and cross

11 a) Any half of the circle coloured
    b) Any quarter of the circle coloured

Notes

Progress Test Charts

Progress Test 1

Q	Topic	✓ or ✗	See page
1	Doubling, Halving, Adding and Subtracting		26
2	Doubling, Halving, Adding and Subtracting		26
3	Place Value		22
4	Solving Number Problems		28
5	Doubling, Halving, Adding and Subtracting		26
6	Place Value		22
7	Numbers and Counting		12
8	Place Value		22
9	Counting More and Less		20
10	Numbers and Counting		12
11	Solving Number Problems		28
12	Place Value		22
13	Doubling, Halving, Adding and Subtracting		26
14	Counting in Steps of 2, 5 and 10 Solving Number Problems		16 28

Progress Test 2

Q	Topic	✓ or ✗	See page
1	Numbers and Counting		12
2	Place Value		22
3	Fractions of Groups		60
4	Doubling, Halving, Adding and Subtracting		26
5	Doubling, Halving, Adding and Subtracting		26
6	What is Multiplication?		40
7	What is Multiplication?		40
8	Counting in Steps of 2, 5 and 10		16
9	Counting More and Less		20
10	Less Than, Greater Than and Equal To		24
11	Numbers and Counting		12
12	Fractions of Groups		60

Progress Test Charts

Progress Test 3

Q	Topic	✓ or ✗	See page
1	Fractions of Groups		60
2	What is Division?		42
3	Solving Multiplication and Division Problems		50
4	Solving Multiplication and Division Problems		50
5	Measuring Weight and Capacity		72
6	Fractions of Groups		60
7	Less Than, Greater Than and Equal To		24
8	Measuring Length and Height		70
9	Measuring Time		76
10	Measuring Weight and Capacity		72
11	Place Value		22
12	What is Division?		42
13	3-D Shapes		86

Progress Test 4

Q	Topic	✓ or ✗	See page
1	Numbers and Counting		12
2	Doubling, Halving, Adding and Subtracting		26
3	Solving Multiplication and Division Problems		50
4	More Doubling, Halving and Dividing		48
5	Left and Right Turns		100
6	What is Multiplication?		40
7	Solving Number Problems Time Problems		28 78
8	Numbers and Counting		12
9	Measuring Length and Height		70
10	Describing Positions		98
11	Halves as Fractions Quarters as Fractions		56 58

What am I doing well in? _____

_____

What do I need to improve? _____

_____